MAMMALS
Primates
Insect-Eaters
AND
Baleen Whales

Robin Kerrod

Facts On File
New York • Oxford

PRIMATES, INSECT-EATERS AND
BALEEN WHALES
The Encyclopedia of the Animal World:
Mammals

Managing Editor: Lionel Bender
Art Editor: Ben White
Designer: Malcolm Smythe
Text Editor: Miles Litvinoff
Assistant Editor: Madeleine Samuel
Project Editor: Graham Bateman
Production: Clive Sparling

Media conversion and typesetting:
Robert and Peter MacDonald

AN EQUINOX BOOK

Planned and produced by:
Equinox (Oxford) Limited,
Musterlin House, Jordan Hill Road,
Oxford OX2 8DP

Prepared by Lionheart Books

Library of Congress
Cataloging-in-Publication Data
Kerrod, Robin.
Mammals: primates, insectivores, and
baleen whales/Robin Kerrod.
p.96, cm.22.5 × 27.5 (The Encyclopedia of
the animal world)
Bibliography: p.1
Includes index.
Summary: Introduces those mammals
whose diet is primarily insects, including the
aardvark, anteater, and bat.

1. Primates – Juvenile literature. 2.
Insectivora – Juvenile literature. 3.
Whales – Juvenile literature. [1.
Insectivores. 2. Mammals.]
I. Title. II. Series.

QL706.2.K47 1988   599-dc19
88-16931

ISBN 0-8160-1961-4

Published in North America by
Facts on File, Inc.,
460 Park Avenue South,
New York, N.Y. 10016

Origination by Alpha Reprographics Ltd,
Harefield, Middx, England

Printed in Italy.

10 9 8 7 6 5 4 3 2 1

FACT PANEL: Key to symbols denoting general features of animals

SYMBOLS WITH NO WORDS

Activity time

- ● Nocturnal
- ◐ Daytime
- ◖ Dawn/Dusk
- ○ All the time

Group size

- ◪ Solitary
- ▥ Pairs
- ◧ Small groups (up to 10)
- ■ Herds/Flocks
- ◢ Variable

Conservation status

- ☠ All species threatened
- ☠ Some species threatened
- No species threatened (no symbol)

SYMBOLS NEXT TO HEADINGS

Habitat

- ◥ General
- ◣ Mountain/Moorland
- ◢ Desert
- ▨ Sea
- ■ Amphibious

- ◿ Tundra
- ◿ Forest/Woodland
- ● Grassland
- ◉ Freshwater

Diet

- ■ Other animals
- ■ Plants
- ◹ Animals and Plants

Breeding

- ◎ Seasonal (at fixed times)
- ◡ Non-seasonal (at any time)

CONTENTS

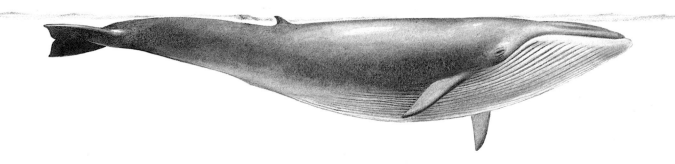

PREFACE

The National Wildlife Federation

For the wildlife of the world, 1936 was a very big year. That's when the National Wildlife Federation formed to help conserve the millions of species of animals and plants that call Earth their home. In trying to do such an important job, the Federation has grown to be the largest conservation group of its kind.

Today, plants and animals face more dangers than ever before. As the human population grows and takes over more and more land, the wild places of the world disappear. As people produce more and more chemicals and cars and other products to make life better for themselves, the environment often becomes worse for wildlife.

But there is some good news. Many animals are better off today than when the National Wildlife Federation began. Alligators, wild turkeys, deer, wood ducks, and others are thriving – thanks to the hard work of everyone who cares about wildlife.

The Federation's number one job has always been education. We teach kids the wonders of nature through *Your Big Backyard* and *Ranger Rick* magazines and our annual National Wildlife Week celebration. We teach grown-ups the importance of a clean environment through *National Wildlife* and *International Wildlife* magazines. And we help teachers teach about wildlife with our environmental education activity series called *Naturescope*.

The National Wildlife Federation is nearly five million people, all working as one. We all know that by helping wildlife, we are also helping ourselves. Together we have helped pass laws that have cleaned up our air and water, protected endangered species, and left grand old forests standing tall.

You can help too. Every time you plant a bush that becomes a home to a butterfly, every time you help clean a lake or river of trash, every time you walk instead of asking for a ride in a car – you are part of the wildlife team.

You are also doing your part by learning all you can about the wildlife of the world. That's why the National Wildlife Federation is happy to help bring you this Encyclopedia. We hope you enjoy it.

Jay D. Hair, President
National Wildlife Federation

INTRODUCTION

The Encyclopedia of the Animal World surveys the main groups and species of animals alive today. Written by a team of specialists, it includes the most current information and the newest ideas on animal behavior and survival. The Encyclopedia looks at how the shape and form of an animal reflect its life-style – the ways in which a creature's size, color, feeding methods and defenses have all evolved in relationship to a particular diet, climate and habitat. Discussed also are the ways in which human activities often disrupt natural ecosystems and threaten the survival of many species.

In this Encyclopedia the animals are grouped on the basis of their body structure and their evolution from common ancestors. Thus, there are single volumes or groups of volumes on mammals, birds, reptiles and amphibians, fish, insects and so on. Within these major categories, the animals are grouped according to their feeding habits or general life-styles. Because there is so much information on the animals in two of these major categories, there are four volumes devoted to mammals (The Small Plant-Eaters; The Hunters; The Large Plant-Eaters; Primates, Insect-Eaters and Baleen Whales) and three to birds (Waterbirds; Aerial Hunters and Flightless Birds; Plant- and Seed-Eaters).

This volume, Mammals – Primates, Insect-Eaters and Baleen Whales, includes entries on monkeys and apes, bats, echidnas, the platypus and Baleen whales. Together they number some 1,550 species. For the most part, they fall into two groups: generalists (those who eat a variety of foods, both plant and animal) and specialists (those whose diet consists only of insects or other small animals that occur in great abundance).

The primates – the lemurs, tarsiers, bush babies, monkeys and apes – are as a group generalists and are our closest relatives. Some form highly sophisticated social relationships, and many are famous for their agility, intelligence and playful behavior.

Many mammals eat insects or other small creatures as a main part of their diet. Some – the shrews, moles and hedgehogs – belong to a specialized group that scientists call the insectivores. But this volume also includes unrelated insect-eating animals such as bats, anteaters and armadillos. Even the Baleen whales (Gray whale, rorquals) are included – not because they eat insects, of course, but because they are adapted to eating small creatures, which they scoop up from the sea in great numbers in their huge mouths. (The other group of whales – the Toothed whales – are dealt with in Volume 2 – The Hunters).

Each article in this Encyclopedia is devoted to an individual species or group of closely related species. The text starts with a short scene-setting story that highlights one or more of the animal's unique features. It then continues with details of the most interesting aspects of the animal's physical features and abilities, diet and feeding behavior, and general life-style. It also covers conservation and the animal's relationships with people.

A fact panel provides easy reference to the main features of distribution (natural, not introductions to other areas by humans), habitat, diet, size, color, pregnancy and birth, and lifespan. (An explanation of the color coded symbols is given on page 2 of the book.) The panel also includes a list of the common and scientific (Latin) names of species mentioned in the main text and photo captions. For species illustrated in major artwork panels but not described elsewhere, the names are given in the caption accompanying the artwork. In such illustrations, all animals are shown to scale; actual dimensions may be found in the text. To help the reader appreciate the size of the animals, in the upper right part of the page at the beginning of an article are scale drawings comparing the size of the species with that of a human being (or of a human foot).

Many species of animal are threatened with extinction as a result of human activities. In this Encyclopedia the following terms are used to show the status of a species as defined by the International Union for the Conservation of Nature and Natural Resources:

Endangered – in danger of extinction unless their habitat is no longer destroyed and they are not hunted by people.

Vulnerable – likely to become endangered in the near future.

Rare – exist in small numbers but neither endangered nor vulnerable at present.

A glossary provides definitions of technical terms used in the book. A common name and scientific (Latin) name index provide easy access to text and illustrations.

LEMURS

It's a stand-off. Two Ring-tailed lemurs face each other across a clearing on the forest floor. Angrily they wave their black-and-white-banded tails, heavily smeared in scent, high above their heads. This is a "stink fight". Scattered behind each of them is a family group, watching tensely. The two groups inhabit the same territory and often come up against each other. But suddenly, as if by a secret signal, the rival groups begin to drift apart.

Travelling through the forests of Madagascar, one will often see ghostly faces peering out through the branches. They are the faces of lemurs, a local word meaning "ghosts." They are not ghosts, of course, but furry, bushy-tailed animals.

Altogether there are over 20 species of lemurs. They all live on the island of Madagascar or on the nearby Comoro Islands.

Lemurs can be seen in almost all of the forests on Madagascar, but different species are found in different regions. The Ring-tailed lemur can be found throughout the island in the deciduous forests. Here most of the trees shed their leaves each year. The Western gentle lemur, on the other hand, is found only in bamboo forests along the western coast. The Brown lesser mouse lemur and the indri live only in the east, the one in forest fringes, the other in the rain forests.

The various species of lemur differ in size, color and way of life. The Ring-tailed lemur is gray, while the Western gentle lemur is brown. Both are about the size of a gray squirrel. By contrast, the Brown lesser mouse lemur is mouse-size, while the indri, which has white fur, is nearly as big as a chimpanzee. Lemurs are primates, like monkeys and apes, but are not so highly developed or intelligent.

LEMURS Lemuridae, Cheirogaleidae, Indriidae, Daubentoniidae (*23 species*)

Habitat: deciduous, rain or bamboo forest.

Diet: mainly vegetarian – leaves, flowers, fruits; some insects.

Breeding: mostly 1 offspring after pregnancy of 4-4½ months.

Size: smallest (mouse lemur): head-tail 10in, weight 1¾ ounces; largest (indri): head-tail 30in, weight 22lb.

Color: brown, white, black and white.

Lifespan: up to 18 years in captivity.

Species mentioned in text:
Aye-aye (*Daubentonia madagascariensis*)
Black lemur (*Lemur macaco*)
Brown lemur (*L. fulvus*)
Brown lesser mouse lemur (*Microcebus rufus*)
Fat-tailed dwarf lemur (*Cheirogaleus medius*)
Indri (*Indri indri*)
Mongoose lemur (*Lemur mongoz*)
Red-fronted lemur (*L. fulvus rufus*)
Ring-tailed lemur (*L. catta*)
Sifakas (*Propithecus verreauxi* and *P. diadema*)
Sportive lemur (*Lepilemur mustelinus*)
Western gentle lemur (*Hapalemur griseus occidentalis*)
Woolly lemur (*Avahi laniger*)

◄An indri in mid-leap showing its unique feature among the lemurs, a stumpy tail.

TYPICAL LEMURS

The Ring-tailed lemur is one of seven species of the so-called typical lemurs. These species are active during the day and feed on leaves, fruit and flowers. This group also includes the Brown, the Red-fronted and the Black lemur. They usually live together in small groups, but occasionally as many as 30 animals may be seen feeding together.

Most typical lemurs remain in the trees nearly all the time. They come down to the ground only where the forest thins out and they cannot leap across the gaps between trees. But the Ring-tailed lemur prefers to travel on all fours down on the ground.

The sense of smell is very important for lemurs. They often stop to smear branches with scent from glands in their body. These scent markings help "signpost" their territory and warn off other lemur groups. Lemurs also smear scent on their tail from glands on their wrists. Waving their scented tail plays an important part in their contests with their rivals.

BAMBOO EATERS

The Sportive lemur and the gentle lemurs behave differently from the typical lemurs. The Sportive lemur becomes active by night and sleeps

▲ Some members of the lemur family
All lemurs have long bushy tails. The muzzle is black and pointed and has sensitive whiskers. The Gray gentle lemur (*Hapalemur griseus griseus*) (1) marks a branch with scent glands on its wrists. The Brown lemur (2) marks its tail with scent. Also seen are the White-fronted lemur (*Lemur fulvus albifrons*) (3), the Ruffed lemur (*Varecia variegata*) (4), and the Sportive lemur (5).

during the day. It also spends much of the time alone.

Sometimes a pair of Sportive lemurs share the same region of forest and come together several times a night to feed and occasionally groom each

other. They feed mainly on leaves. The gentle lemurs, though, have a limited diet, feeding only on bamboo shoots and reeds. Sometimes seen in small groups, they are most active in the morning and early afternoon. They use their hands a lot when they eat, to get at the tender inner parts of bamboo shoots and push them into the mouth.

THE SMALLEST PRIMATES

Dwarf and mouse lemurs are the smallest lemurs, and indeed are the smallest of all primates. The smallest mouse lemur weighs less than 2 ounces even when fully grown. Both the dwarf and mouse lemurs are nocturnal. They have a much more varied diet than the other lemurs, eating beetles and other insects, as well as fruit, leaves and gum that comes out of the bark of trees.

Dwarf and mouse lemurs live alone for most of the time, except in the mating season. Unlike most other lemurs, they usually give birth to two or three offspring.

When food is plentiful, in the wet season, mouse lemurs build up fat in their rump and tail. This store helps them survive in the drier months when food becomes scarce. Dwarf lemurs store fat in the same way. But

▲ ▼Among the lemurs, the Ring-tailed lemur is easiest to recognize. No other lemur has a tail quite like it. In a "stink fight" the lemur rubs scent on its tail, then shakes it back and forth to waft the smell towards its rival.

they become dormant, or sleep, through the dry months.

LEAPS, HOOTS AND RATTLES

Most lemurs move through the leafy branches of the forest by leaping. But the name "leaping lemurs" is usually given to the large lemurs of the indri family. They include the indri itself, the Woolly lemur and the sifakas.

These animals are truly master-leapers. They launch themselves through the air in an upright position, with arms outstretched, from tree trunk to tree trunk. The indri has been seen to make leaps of up to 33ft. It also has an odd way of walking on the few occasions it descends to the ground. It hops on its hind legs, holding its arms in the air.

The leaping lemurs live together in small groups. They mark their territory with their scent and give out frequent loud howls. They use different calls to warn of danger. The indri hoots and roars; sifakas make a noise like a New Year's Eve noisemaker.

LEMURS AT RISK

All the lemurs live and feed in the forests, and they are affected when the trees are cut down for timber or to increase farmland. Some species are already scarce and are found only in small areas of Madagascar. Most at risk at present is the aye-aye, which lives in the rain forests of the eastern coast.

The gentle lemurs are also becoming scarce. This is mainly because their diet of bamboo shoots and reeds is found in only a few places. The other lemurs stand a better chance of surviving because they eat a variety of food. Sifakas suffer more directly from human contact. Local people trap and shoot them for food.

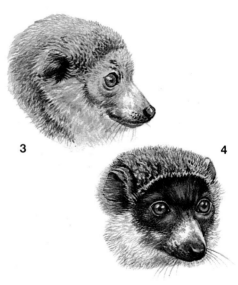

▲With some lemurs the male and female look different, as with the male (1) and female (2) Black lemur, where only the male is really black. The male Mongoose lemur (3) has lighter coloring than the female (4).

◀The Fat-tailed dwarf lemur, so called because it stores fat in its tail, is a slow-moving animal.

TARSIERS

Scuttling through the wet leaves on the forest floor, a cockroach searches for food. It is unaware that two huge eyes are following its every move. A tarsier is out hunting. When the cockroach gets within reach, the tarsier leaps from the trees and pins it to the ground. A quick bite with razor-sharp teeth, and the insect is dead. The successful hunter returns to its perch and devours its prey.

Tarsiers are mammals that grow to the size of a rat. They have a thick, furry, velvet-soft coat. All three species of tarsier look very much alike, with only slight differences in the tail. They are tree-dwellers of the tropical rain forests.

Like the lemurs and bush babies, tarsiers are primates. They first appeared on Earth at least 50 million years ago, before the simians, the monkeys and apes. For this reason they are called prosimians.

ISLAND-DWELLERS

Tarsiers are found in Borneo, Indonesia and the Philippines, all islands that straddle the equator in South-east Asia. The Western tarsier lives in Borneo and Sumatra. The Spectral tarsier lives in Sulawesi. It is also called the Celebes tarsier after the old name for this island. Both species are found in very large numbers.

The Philippine tarsier is much rarer. It inhabits islands in the south of the Philippines and suffers because of tree-cutting operations there. Logging kills many tarsiers every year and also destroys their habitat. They need to be protected if they are to survive.

ENORMOUS EYES

The most striking feature of the tarsier is the eyes. They are huge. In the Western species, each eye weighs more than the brain. The size of the eyes tells us that tarsiers are nocturnal creatures, active at night.

The eyes look forward but cannot move much in their sockets. To look around, the tarsier has to move the whole head. It can swivel its head through a wide angle, like an owl.

Other unusual features of the tarsier's body are the hind legs and tail. The animal has very long hind legs indeed – twice as long as the rest of the body. These long limbs make the

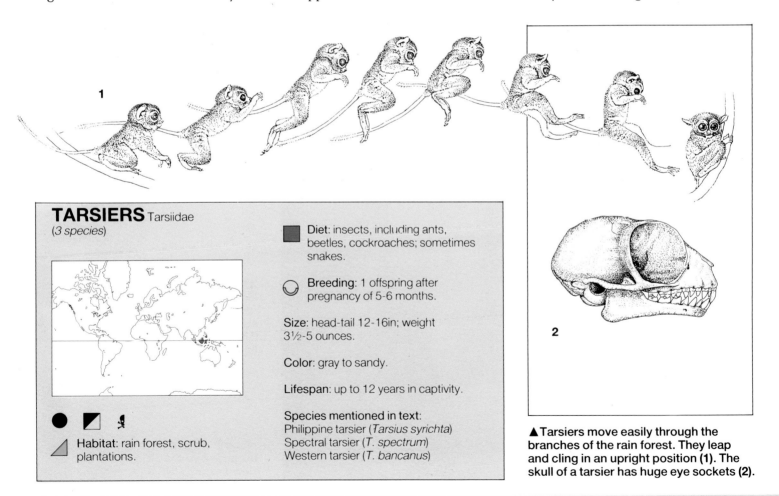

TARSIERS Tarsiidae
(*3 species*)

Diet: insects, including ants, beetles, cockroaches; sometimes snakes.

Breeding: 1 offspring after pregnancy of 5-6 months.

Size: head-tail 12-16in; weight 3½-5 ounces.

Color: gray to sandy.

Lifespan: up to 12 years in captivity.

Species mentioned in text:
Philippine tarsier (*Tarsius syrichta*)
Spectral tarsier (*T. spectrum*)
Western tarsier (*T. bancanus*)

Habitat: rain forest, scrub, plantations.

▲Tarsiers move easily through the branches of the rain forest. They leap and cling in an upright position (**1**). The skull of a tarsier has huge eye sockets (**2**).

tarsier an outstanding leaper. They propel the animal into the air like a powerful spring.

The tarsier's tail is as long as its hind legs and has a tuft of hair at the end. When leaping, the tail is extended backwards to help the animal keep its balance.

ENORMOUS HANDS

Tarsiers have almost human-like hands on the front limbs. They have long, slender fingers and finger-nails. The tips of the fingers are covered in pads, which help them cling to the branches and tree trunks they leap on. Tarsiers also use their fingers as a kind of "cage" to trap swift-flying night insects for food.

The toes on the hind legs are similar to the fingers, except that the second and third toes have claws instead of nails. They are called toilet claws, because the tarsier uses them when grooming.

DUETS

Unlike some of their relatives, tarsiers are not noisy creatures. But they often call from around their sleeping area in the middle of their home territory. The male and female Sulawesi tarsier make beautiful music as they sing duets together in high-pitched voices.

In their courtship, the male and female Western tarsier call quietly to one another as they scamper around. When left behind by their mother out hunting, the young tarsiers call to her with soft clicking noises or whistles.

▼The huge eyes of the Philippine tarsier help it see well even on the darkest night.

▲The Western tarsier, like its relatives, has very long hind legs. They provide the power to propel the animal through the air when leaping.

BUSH BABIES AND LORISES

In a dense forest in Gabon, late in the evening, an urgent call suddenly pierces the darkness: "Ngok! Nogkoué!" It is the alarm call of a bush baby high in the tree canopy. Some large animal is passing near by, and the bush baby senses danger.

BUSH BABIES AND LORISES
Lorisidae (*10 species*)

 ■

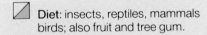 Habitat: tropical rain forest, dry forest, savannah with trees.

Diet: insects, reptiles, mammals birds; also fruit and tree gum.

Breeding: usually 1 offspring after pregnancy of 4-6 months.

Size: smallest (Dwarf bush baby): head-tail 12in, weight 2 ounces; largest (Thick-tailed bush baby): head-tail 30in, weight 2½lb.

Color: gray to reddish-brown.

Lifespan: up to 15 years in captivity.

Species mentioned in text:
Allen's bush baby (*Galago alleni*)
Dwarf bush baby (*G. demidovii*)
Golden potto, or angwantibo (*Arctocebus calabarensis*)
Lesser bush baby (*Galago senegalensis*)
Needle-clawed bush baby (*G. elegantulus* W. Africa and *G. inustus* C. Africa)
Potto (*Perodicticus potto*)
Slender loris (*Loris tardigradus*)
Slow loris (*Nycticebus coucang*)

Bush babies live only in Africa, but are found widely there. Allen's bush baby lives in the tropical rain forests of West Africa. So do the Needle-clawed bush baby and the Dwarf bush baby. The latter can also be found much farther afield, on the coast of Kenya in the east. The most widespread species of all, however, is the Lesser bush baby. This can be found in wet and dry forests and open woodland from Senegal to Angola in the west and from Ethiopia to Mozambique in the north and east.

Pottos are also found widely in the West African rain forests. Lorises, however, live in Asia. The Slender loris is found in India and Sri Lanka, the Slow loris over a vast area from Bangladesh in the north to Indonesia in the south.

FAST MOVERS
Bush babies are well named. They look appealing and have sad cries. They are also sometimes called galagos after their scientific genus name. Like the lorises and pottos, they are primates and are related to the lemurs of Madagascar.

Bush babies, like all creatures of the night, have large eyes. They also have large ears with which they can detect the slightest sounds. They use their sensitive ears when hunting for insects such as moths, which make up much of their diet. When a moth flies by, they clamp their feet around a branch. Stretching out their body, they snatch the moth out of the air with one or both hands.

The arms of bush babies are quite short, while the hind legs are long and muscular. This tells us that they should be good at leaping, and so they are. They can move rapidly through the leafy canopy of the forest and are usually able to escape from animals that hunt them. Their long bushy tail helps steady them as they jump.

In Gabon, Allen's bush baby is known as Ngok or Nogkoué, after the alarm call it makes when a dangerous animal such as a leopard passes nearby. Bush babies are in general very vocal during the night as they spread out to feed. They call to keep in contact with other members of their group. In the early morning a rallying call brings them together so that they can sleep as a group.

SLOW CLIMBERS
Pottos and lorises are close relatives of bush babies. But instead of being swift jumpers they are slow climbers. They move as though in slow motion through the thick forests they live in. This makes them very difficult to see. When they are scared by a sudden sound, they freeze. They can remain in a fixed position for hours if need be until danger has passed.

Pottos and lorises have large eyes because they are nocturnal, like bush babies. But, unlike bush babies, all four of their limbs are about the same length, and they have only a stumpy tail. They feed on slow-moving creatures, such as caterpillars and beetles. They hunt out their prey with their nose, for they have a well-developed sense of smell.

Being slow movers, pottos and lorises could become easy prey themselves for hunters like the civet. But they can defend themselves quite well. The potto uses a hump of thick skin on its shoulder as a shield to defend itself. The Golden potto rolls itself up into a ball. Both pottos can bite fiercely when attacked.

NOT ENDANGERED
Because of their small size and nocturnal habits, bush babies, lorises and pottos are not hunted as much as other primates in the countries where they live. In fact, few people have seen these animals in the wild and, except in those areas where their forest homes are being destroyed, there is little that threatens their survival.

▶Moving about in the trees. Like the other bush babies, the Thick-tailed bush baby (*Galago crassicaudatus*) (1) travels rapidly by leaping. Its relatives, the lorises and pottos, are climbers, moving slowly along the branches. They include the Slow loris (2), the Slender loris (3), the potto (4) and the Golden potto (5). Their hands and feet have a strong pincer-like grip.

MARMOSETS AND TAMARINS

It is just after dawn in a coastal forest of Brazil. Resting in a cosy hole in one of the tall trees is a family of Golden lion tamarins. All at once, the peace is shattered by the roar of a chain saw. Startled, the mother and father, each carrying one of their twin babies, scamper around in panic, then head into the forest, away from the noise. But worse is to come. In their headlong flight from the din, they don't hear the poachers. Two shots ring out, and the parents are dead. The poachers snatch the babies, still clinging in terror to their parents' bodies. The babies will fetch a good price at market.

Poaching and the destruction of forests have made the Golden lion tamarin almost extinct in the wild. Fewer than 100 may remain. The same is true for a close relative, the Golden-rumped lion tamarin. Other tamarins and some species of marmosets are also becoming rare because of human interference.

Marmosets and tamarins are the smallest of the monkeys that live in the Americas. They mostly inhabit the thick forests of the River Amazon basin in Brazil. Some live in more open savannah-type country in Paraguay and Bolivia. Others are found as far north as Costa Rica in Central America.

Because of their attractive appearance, marmosets and tamarins have long been in demand as pets and for zoos. They have a fine, silky, often colorful coat. Some have a mane, some a crest and others the most splendid drooping moustaches.

The rare Golden lion tamarin's bright gold mane, like a male lion's, makes it one of the most striking creatures on Earth. The Cotton-top tamarin of Columbia has a pure white crest that cascades to its shoulders. Among the moustached tamarins, the Emperor tamarin has an impressive white beard.

A STICKY DIET

Both marmosets and tamarins have a varied diet, feeding on plants, insects and other animals. They also feed on the gum oozing from trees. This often occurs when the trees are bored into by insects. Marmosets, however, do not need to rely on insects to hole the trees. They can gouge out holes themselves. When they want to make a hole, they sink their long upper incisor teeth into the bark and then gouge upwards with their lower ones, which act like a chisel.

The Pygmy marmoset feeds on gum more than the other marmosets. Like them, it has claws on all its fingers and toes, except for the big toes. This helps it cling to the tree trunk while feeding.

FAMILY GROUPS

Marmosets and tamarins go around in small family groups of mother, father and their offspring. Only one female in the group breeds, even though there may be several others present. She gives birth usually twice a year and frequently to twins.

When it comes to looking after the baby, every member of the group helps out, including the father. All the group also help to feed the mother, as well as the infants and other group members that are carrying them.

▶ **Marmosets and tamarins** Goeldi's monkey (*Callimico goeldii*) (1) and the Black-tailed marmoset (*Callithrix argentata melanura*) (2) are shown in offensive pose. Geoffroy's tamarin (*Saguinus geoffroyi*) (3), the Red-chested moustached tamarin (*Saguinus labiatus*) (4) and the Saddle-back tamarin (*Saguinus fuscicollis*) (5) are seen scent-marking. The Golden-rumped lion tamarin (6) has a long flowing mane, while the Tassel-ear marmoset (*Callithrix humeralifer intermedius*) (7) has a red muzzle. The smallest marmoset, the Pygmy (8), gouges a tree for gum.

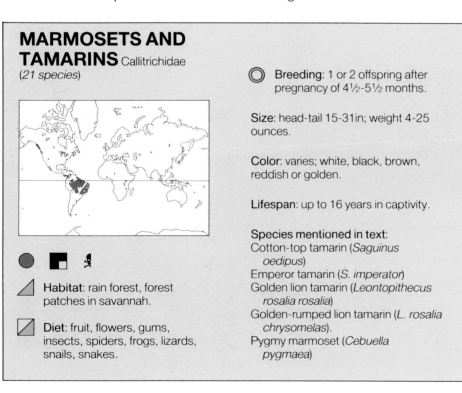

MARMOSETS AND TAMARINS Callitrichidae
(*21 species*)

⊙ Breeding: 1 or 2 offspring after pregnancy of 4½-5½ months.

Size: head-tail 15-31in; weight 4-25 ounces.

Color: varies; white, black, brown, reddish or golden.

Lifespan: up to 16 years in captivity.

Species mentioned in text:
Cotton-top tamarin (*Saguinus oedipus*)
Emperor tamarin (*S. imperator*)
Golden lion tamarin (*Leontopithecus rosalia rosalia*)
Golden-rumped lion tamarin (*L. rosalia chrysomelas*).
Pygmy marmoset (*Cebuella pygmaea*)

◢ Habitat: rain forest, forest patches in savannah.

◢ Diet: fruit, flowers, gums, insects, spiders, frogs, lizards, snails, snakes.

CAPUCHIN MONKEYS

The Sun sets over the vast Amazon jungle. For 50 species of monkey it is time to sleep. But in thick tangled vines and holes in trees another monkey species is stirring. As darkness falls, Night monkeys spread out, foraging almost silently for food. Then the sky begins to lighten, as a brilliant full Moon rises. The hooting starts – it is a group of male Night monkeys. They are declaring their territory and warning off other males.

The tropical forests of the Amazon basin are the home of the Night monkey and of many other species of the capuchin family. These include not only the capuchin monkeys themselves, but also Squirrel monkeys, titis, sakis, uakaris and spider monkeys. The capuchin family are often called cebids, after the Latin family name.

The Night monkey and Squirrel monkey are found throughout tropical South America and in Central America. Other monkeys of the capuchin family have a considerably narrower range. The Yellow-tailed woolly monkey, for example, is found only in one mountainous region in Peru.

Capuchin monkeys are named after Capuchin monks. This is because they have a cap of hair on the crown of the head which looks like a monk's cowl or hood. The monkeys are among the most intelligent of all the primates and are easily trained as pets and to perform tricks.

COMPETITION FOR FOOD

Several different species of monkeys often live in the same region of forest, sharing the food the forest provides.

▶Like its parents, this baby Smoky woolly monkey is covered from head to toe in long thick fur. When older, its coat may turn gray or brown, but its head will remain dark.

CAPUCHIN MONKEYS
Cebidae (*30 species*)

● ■ ⚹

◢ Habitat: tropical and subtropical evergreen forest.

◸ Diet: fruit, roots, leaves, insects, small mammals, snails, spiders.

◎ Breeding: 1 offspring after pregnancy of 4-7½ months.

Size: smallest (Squirrel monkey): head-tail 25in, weight 21 ounces; largest (Woolly spider monkey): head-tail 55in, weight 26lb.

Color: varies; white, yellow, red-brown, black, often with patterning around head.

Lifespan: up to 25 years.

Species mentioned in text:
Brown capuchin (*Cebus apella*)
Mexican black howler (*Alouatta villosa*)
Night monkey (*Aotus trivirgatus*)
Smoky, or Humboldt's woolly monkey
  (*Lagothrix lagotricha*)
Squirrel monkey (*Saimiri sciureus*)
White-faced saki (*Pithecia pithecia*)
Woolly spider monkey or muriqui
  (*Brachyteles arachnoides*)
Yellow-tailed woolly monkey (*Lagothrix flavicauda*)

▲Capuchin monkeys moving through
the trees The bushy-tailed White-faced
saki (1) and stumpy-tailed Red uakari
(*Cacajao rubicundus*) (2) climb through
the branches on all fours. The Dusky titi
(*Callicebus moloch*) (3) and Squirrel
monkey (4) travel mainly by leaping. The
Black howler monkey (*Alouatta caraya*)
(5), Black-handed spider monkey (*Ateles
geoffroyi*) (6) and Smoky woolly monkey
(7) use their tails as well as their hands.
The animals shown are all females. The
males are slightly bigger.

►The Woolly spider monkey is the largest and most ape-like of the New World
monkeys. It swings by its tail and arms.

This can sometimes lead to as many as five different species feeding on the same tree. As is to be expected, there is usually a fight to see which one will have the best of the food.

The big monkeys usually win, but not always. The Squirrel monkey relies on safety in numbers, going about in large groups of 30 to 40. Such numbers are usually more than a

match for the small groups of large monkeys that might threaten them. The Night monkey has overcome the competition by adapting to a nocturnal life. It feeds while its rivals sleep.

Other small monkeys have taken to a more specialized diet. The titis are able to eat fruit while it is still green, which the big monkeys will not touch. The bearded sakis can open the seeds of fruits and eat what is inside.

LEAPERS AND SWINGERS
Cebids spend almost all their time in the trees. Only occasionally do they drop to the ground to play, search for food or travel between stands of trees. Up in the trees they move about by leaping, climbing and swinging on their arms. The leapers, such as the Squirrel monkey, have long powerful hind legs for launching themselves into the air.

The "swingers," such as the spider monkeys, have long arms and move hand over hand below the branches. They have specially developed shoulder-joints that help them swivel easily. Their most interesting feature is the tail, which is long and flexible. Known as a prehensile tail, the monkeys use it as a fifth limb to grip branches and to steady themselves. They may even hang from it, leaving their hands free to reach food.

BREEDING PAIRS
Most of the smaller cebids, such as the Night monkey, the titis and the sakis, live in small family groups, based on a male and female pair. The pair stay together, raising one infant each year. The young stays with the parents until after the next offspring arrives.

The small Squirrel monkey, living in much larger groups, is an exception. Each group contains several breeding males and females. Vicious fights break out in the breeding season between the males. In some of the larger species, the males keep a harem of several females.

SAFETY IN NUMBERS
Spider monkeys and capuchin monkeys also often live together in groups of up to 20. Monkeys belonging to a large group are less likely to suffer from attacks by predators. These include a variety of hawks and eagles. When a group is large, there are more eyes and ears to watch and listen and to raise the alarm when danger threatens.

A large group size also helps in the constant search for food. Individual monkeys in the group may know of different food trees and when they are

▲Young Squirrel monkeys spend much of their time playing. They wrestle (1) and spar (2) with one another, explore holes in trees (3), stalk birds (4) and try to catch insects (5). All these activities help prepare them for adult life.

in fruit. A large group can also send out more "scouts" to look for new fruit trees. When a member of a Brown capuchin group finds a new tree, it lets out a loud whistle to call the others.

Another advantage is that a large group can usually chase monkeys of smaller groups away from the best food trees. It can also defend the trees better.

Sometimes different groups live peacefully together. Groups of Squirrel monkeys do not usually fight even when they feed on the same trees. They may also join up with groups of Brown capuchins to feed, again without any fighting.

PECKING ORDER

Life in most groups is well organized. Most activities center around looking for food and raising young. There is nearly always a "pecking order" among the members of the group. The leader is usually a dominant male. The social position of other group members depends on how well they get along with him.

The pecking order is often important at feeding time. In a Brown capuchin group, the "boss" and his favorites usually feed first. He positions himself inside a ring of other monkeys so as to be safer from predators. The monkeys not in his favor may have to wait until he and his select group have finished eating and moved on.

REARING INFANTS

Most members of a group share in raising the young. A new mother is helped by females that do not have offspring of their own. These monkeys help carry the young and also "baby-sit" while the mother goes off to feed.

In some species, including the Squirrel monkey, the males play little part in family life. But in species which live in small family groups the males

▲The Squirrel monkey sometimes eats flowers as well as its usual diet of fruit and insects.

▼A family group of titi monkeys. When resting, titis huddle together and entwine their tails (1). The father often grooms the young (2).

A pair of White-faced sakis. Only the male (left) has a white face. They will probably spend most of their life together in a small family group.

help out. In the small family groups of the titi monkeys, the young often spend more time with their father than with their mother. The father carries the youngest most of the time and spends hours grooming it.

THE SUCCESSFUL HOWLERS

Most monkey groups take part in the dawn chorus of the jungle. They call at other times during the day to announce where they are and to warn off other groups near by. The loudest monkeys by far are the six species of howler monkey.

By passing air through an enlarged bone in their throat, they can let out a very penetrating howl. It can carry for over half a mile, even through thick forest. The loudness of their howl helps individuals and groups keep in touch with each other and has helped make them successful as a species. Howlers have the widest range of all the primates in the New World. The Mexican black howler is found as far north as Yucatan in Mexico, the Black howler as far south as Argentina.

Another reason for the howlers'

success is that when fruit and flowers are not available they can eat leaves. They can live on leaves for weeks, something most other species of monkey would find difficult.

Howlers are able to survive on leaves because of their modified gut, in which the leaves are digested. They help themselves by choosing tender young leaves which break down more quickly. However, leaves are not very nutritious and do not provide much energy. As a result, howlers are quite slow moving animals and spend half the day resting or sleeping.

LONG-TERM PROSPECTS

More than 10 species of Capuchin monkeys are endangered. In the Amazon basin, spider monkeys and woolly monkeys have been shot for food over much of their range. One species is now confined to just seven small areas of forest and its numbers have fallen to less than 250. However, species such as the Night monkey and Squirrel monkey can adapt quite easily to a loss and change of habitat. If more of their forest homes are protected, their survival can be assured.

▲Looking almost like a parrot here, this male Night monkey is looking for a mate. He may travel more than 3 miles on his nightly travels.

MANGABEYS AND GUENONS Cercopithecidae
(*23 species*)

Habitat: mostly rain forest, also other forest and savannah.

Diet: fruit, seeds, leaves, flowers, crops, insects.

Breeding: 1 offspring after pregnancy of 5-6 months.

Size: varies; head-body 18½-26in; weight 7¼lb female, 10lb male (Vervet monkey).

Color: mainly black, brown, gray or white, often with different colored crown, collar, cheek patches or moustache.

Lifespan: up to 31 years.

Species mentioned in text:
Agile mangabey (*Cercocebus galeritus*)
Black mangabey (*C. aterrimus*)
Gray-cheeked mangabey (*C. albigena*)
Patas monkey (*Erythrocebus patas*)
Red-eared monkey (*Cercopithecus erythrotis*)
Vervet (*C. aethiops*)
White mangabey (*Cercocebus torquatus*)

A small group of Red-eared monkeys are feeding on succulent shoots in a favorite tree. The monkeys are aware of the dangers of sudden attack by their arch enemies, the eagles. So they are stuffing their multi-colored faces with the shoots they pick. They cram them, after scarcely a chew, into their cheek pouches with the back of their hands. When their cheek pouches are full, they scamper away to the safety of thicker foliage to eat at leisure.

▼The Moustached monkey (*Cerco-pithecus cephus*) **(1)** a guenon living in the rain forests of West/Central Africa, it is so called because of the white "V" on its face. Allen's swamp monkey (*Alleno-pithecus nigroviridis*) **(2)** inhabits the forest swamps of Congo and Zaire. Its diet includes shrimp and fish. The Gray-cheeked mangabey **(3)** belongs to the tropical evergreen forests from Cameroon to Uganda.

◀A troop of vervets foraging for food. Their varied diet includes leaves, fruit, insects, eggs, nestlings and small rodents.

Most species of the more colorfully coated guenons stay up in the trees. One of the exceptions is the Patas monkey, which is the biggest of the guenons. It has very long legs that can carry it at speeds of up to 33mph. This makes it the fastest runner of all primates.

MOTHERS AND DAUGHTERS
Mangabeys and guenons move about in groups of usually about 15, but sometimes more. In most species the group consists mainly of females, with only one or two males. This is because when a pair of monkeys breeds, the female offspring stay with the mother as they grow up, while the young males leave. The daughters may set up their own female group within the same monkey troop. The family link through the mother is called the matriline.

As with most monkey species, mangabeys and guenon groups are noisy. They join in the general dawn chorus of the forest and also call through the day to warn other groups to keep out of their territory. The most distinctive of all the calls is that of the male Gray-cheeked mangabey. Its low-pitched "whoop-gobble" can carry for more than half a mile through the thick tropical forest.

MIXED FORTUNES
All forest-living monkeys are to some extent endangered because their habitat is being destroyed rapidly and over wide areas to make room for farmland and roads or to provide timber and wood pulp. Many species of mangabeys and guenons are further threatened by hunting and by diseases. The vervet, though, which lives mainly in savannah, and has a varied diet, is widespread and in places common.

The Red-eared monkey is an attractive species, with a brown coat, gray limbs and a blue, red and yellow face. Its food-cramming behavior is typical of the mangabeys and guenons, all species of which have cheek pouches. These so-called Old World (Africa and Asia) monkeys differ in this respect from the New World monkeys of Central and South America.

The two sets of monkeys also have a different shaped nose. The New World monkey has a broad nose with wide-open nostrils which face outwards. The Old World monkey has a much narrower nose, with nostrils close together and pointing downwards. Another difference between the two is in their tails. None of the Old World monkeys has a gripping or prehensile tail.

FASTEST PRIMATES
All the mangabeys live in thick canopy forests. They are closely related to baboons and are sometimes called long-tailed baboons. While the Gray-cheeked and the Black mangabey spend most of their time in the tree-tops, the Agile and the White mangabey prefer to stay on the ground, like baboons.

MACAQUES

One of Gibraltar's two troops of Barbary apes are resting quietly on a rocky hillside. They are huddled together in small groups of males, females and babies. Trouble seems to be brewing in one group as one male threatens another. But the threatened male doesn't fight. Instead he picks up one of the babies by the leg and offers it upside-down to the other male. This gesture quickly restores peace and both males start lip-smacking and teeth-chattering over the baby.

MACAQUES
Cercopithecidae (*15 species*)

Habitat: lowland and highland forest, scrubland, cliffs, swamps.

Diet: fruit, leaves, crops, insects.

Breeding: 1 offspring after pregnancy of about 5 months.

Size: varies; head-body 20-28in; weight 11lb female, 17½lb male (Stump-tailed macaque).

Color: mostly light or dark brown or gray; sometimes with a colored rump.

Lifespan: up to 30 years.

Species mentioned in text:
Barbary ape or macaque (*Macaca sylvanus*)
Crab-eating macaque (*M. fascicularis*)
Japanese macaque (*M. fuscata*)
Père David's macaque (*M. thibetana*)
Rhesus monkey or macaque (*M. mulatta*)

The so-called Barbary apes are not in fact apes, but macaques. In North Africa they inhabit the high forests, scrubland and cliffs of the Atlas mountains. They were brought to Gibraltar by British soldiers in 1740.

Barbary apes are hardy creatures. Their thick fur helps them survive in the cold winters of the mountains. Lip-smacking and teeth-chattering are normal behavior among males, who play an important part in looking after babies and infants within or outside their own group. They protect them when danger threatens and carry and groom them.

MACAQUES OF ASIA
Other macaques are hardy mountain dwellers too, living in Asia. They include the Japanese macaque, Père David's macaque of Tibet and China and the Rhesus monkey, which ranges all through the foothills of the Himalayas, from Afghanistan to China.

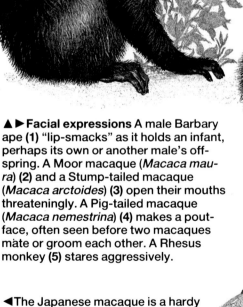

More species of macaques live in the rain forests of the tropics, in Borneo, the Philippines and Indonesia. Among these tropical species is the Crab-eating macaque, which makes its home in swamps and coastal forests and does indeed sometimes eat crabs.

USED FOR SCIENCE

The Rhesus monkey is one of the most wide-ranging and most plentiful of the macaques. It is also the one most widely studied. For many years it has been the monkey most used in laboratory experiments for medical research.

Research scientists use the Rhesus monkey to test the effectiveness of drugs against diseases. (Like its close relatives, the mangabeys, guenons and baboons, the Rhesus monkey can catch many human diseases, including tuberculosis and yellow fever.) This helps scientists discover how the drugs will affect human beings.

Once it was thought that trapping of

Rhesus monkeys for export for medical research might one day wipe out the species. But this now seems unlikely, because many laboratory animals are bred in captivity, and research using animals has declined.

UNDER THREAT

One major threat these days comes from farmers. In some regions Rhesus monkeys and other macaques live on the forest edges and often go into the cultivated fields to feed on the crops. When a troop of up to 70 animals gets to work, the crops are devastated. So the farmers go hunting to reduce the monkey population.

▲▶Facial expressions A male Barbary ape (1) "lip-smacks" as it holds an infant, perhaps its own or another male's off-spring. A Moor macaque (*Macaca maura*) (2) and a Stump-tailed macaque (*Macaca arctoides*) (3) open their mouths threateningly. A Pig-tailed macaque (*Macaca nemestrina*) (4) makes a pout-face, often seen before two macaques mate or groom each other. A Rhesus monkey (5) stares aggressively.

◀The Japanese macaque is a hardy species. Its thick shaggy coat keeps it warm in the cold snowy winters of northern Japan.

BABOONS

A group of about 30 Hamadryas baboons are picking their way through the stony desert landscape. A male near the front of the ragged line looks round to see if the rest of the group is keeping up. He sees one of the females lagging behind and dashes down the line towards her. Seeing him coming, she hurries to catch up. But it is too late. He sinks his teeth into the back of her neck and shakes her angrily. Squealing, she follows him closely back into line.

The Hamadryas baboon lives in very large troops, sometimes of more than 200 animals. A small group of 30 may form a clan of perhaps three families. A number of clans group into a band of 80 or 90 animals, and several bands join together to form the troop.

The smallest group, the family unit, is a harem, led by a male. The leader is followed by his female mate, their daughters and a few male "hangers on". Some of these quietly join the clan to court the young females and in time attract them away to start a

BABOONS Cercopithecidae
(6 species)

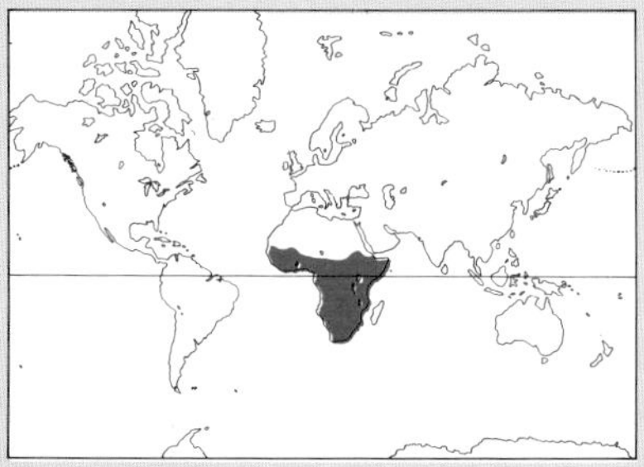

Size: smallest (Savannah baboon): head-body 22-30in, weight up to 31lb female, 46-55lb male; largest (drill): head-body 28in, weight up to 110lb.

Color: tinged grays and browns, colored rump.

Lifespan: up to 30 years.

Species mentioned in text:
Hamadryas baboon (*Papio hamadryas*)
Mandrill (*P. sphinx*)
Savannah, Chacma, Olive, Yellow baboon (*P. cynocephalus*)

Habitat: savannah, woodland, rain forest, desert.

Diet: grass, roots, fruit, seeds, insects, other small animals.

Breeding: 1 offspring after pregnancy of about 6 months.

separate family of their own.

Another species, the Savannah baboon, also lives in huge groups. Savannah baboon families include several adult males rather than just one.

SAVANNAH TO HIGHLANDS

Baboons are the largest of the monkeys. They live almost every-

1

2

3

26

where in Africa where there is water to drink.

The Savannah or Common baboon is widespread in the grasslands and bush and along forest edges from Ethiopia to South Africa. There are three different forms of the Savannah baboon, each from a different region. They are recognizable by the color of their coat. The Yellow baboon (its coat is yellowish-gray) lives in lowland East and Central Africa; the Olive baboon (olive green-gray) in the East African highlands, and the Chacma baboon (with a dark gray coat) in southern Africa.

The Hamadryas baboon lives in Ethiopia and neighboring Somalia and, across the Red Sea, in Saudi Arabia and South Yemen. It is found in rocky desert areas of scattered grass and thorn bush.

COLORFUL FEATURES
Baboons have a naked face and a muzzle rather like that of a dog. Males and females can often be recognized by their coat. The adult male Hamadryas baboon, for example, has long silvery-gray hair, forming a kind of cape over its shoulders. The female's coat is brown.

One can also identify the sexes by the color of the face. Females have a black face, males a bright red one. The males also have a distinctive bright red rump.

The mandrill is the most colorful among baboon males. Its face is marked red and blue, and its bare rump is blue to purple. The female has similar but duller coloring and is only half the male's size.

When they are ready to mate, adult female baboons develop swellings on their rump and thighs. Each individual has a characteristic pattern of swellings.

◀Drills and baboons Red-and-blue-faced mandrill (1) Drill (*Papio leucophaeus*) (2) Gelada (*Theropithecus gelada*) (3) showing bare patches on its neck and chest. Hamadryas baboon (4), with red naked skin on its face and rump. Guinea baboon (*Papio papio*) (5). Olive baboon (6), a form of the Savannah or Common baboon, of highland East Africa, shown with its dead prey, a hare. Chacma baboon (7) another form of the Savannah baboon, of southern Africa. Each example is of the adult male.

The female Proboscis monkey is ready to mate and pads up to a male to which she is attracted. She tries to catch his eye, pursing her lips with the mouth closed. At first the male ignores her, then he returns her glance. She shakes her head, but this means "Yes" not "No". The male then pouts his lips and begins to mate with her. Now pouting too, she continues to shake her head. Instinctively she has timed their mating so that their young will be born when there is the greatest supply of food around, in roughly six months time.

The colobus and leaf-eating monkeys of Asia and Africa are often known as the colobines. They are long-tailed monkeys, which spend most of their time in the trees. They belong to the same scientific family as guenons, mangabeys, macaques and baboons, although there are differences between all these close relatives.

Colobines are more slender. They have no cheek pouches, and their teeth are different too. They also have large salivary glands, and a larger stomach, with two or more chambers. They need extra saliva and a more complicated stomach to help them digest leaves, which are a major part of their diet.

In the stomach, bacteria break down the cellulose in the leaves into sugar. They can also break down poisons. Because of this, the Hanuman langur can eat fruits containing the deadly poison strychnine without coming to any harm.

AMONG THE MANGROVES

The colobines are widespread in Asia. The Proboscis monkey lives on the island of Borneo, in the lowland rain forests and among the mangroves along the river banks. It is quite a good swimmer. The male, with its long fleshy nose, is a hefty animal weighing as much as 50lb. The female is only about half this weight.

Borneo is also the home of five more species of colobine, including the Grizzled sureli and the Silvered leaf monkey. The surelis belong to a genus of monkey called *Presbytis*, meaning "old woman". The name was given to them because their faces are so wrinkled and wizened. The seven species of sureli are found only in the Malay peninsula, Borneo, Sumatra and other Indonesian islands.

The Silvered leaf monkey has a feature common to all leaf monkeys and langurs. It has ridges on its face that make it look as if it is raising its eyebrows.

Other species of leaf monkeys are found in India, Sri Lanka, Bangladesh, Vietnam and China. China is also the home of three snub-nosed monkey species, all of which are becoming rare. They include the attractive Golden monkey. Its coat is tinged a rich orange, its muzzle is white, and its

COLOBUS AND LEAF MONKEYS Cercopithecidae (*37 species*)

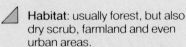

Habitat: usually forest, but also dry scrub, farmland and even urban areas.

Diet: leaves, fruit, buds, flowers, seeds, crops.

Breeding: 1 offspring after pregnancy of 4½-7 months.

Size: smallest (Olive colobus): head-body 17in, weight 6¼lb; largest (Hanuman langur): head-body 43in, weight 53lb.

Color: black, brown, gray, orange.

Lifespan: up to about 20 years.

Species mentioned in text:
Golden leaf monkey (*Semnopithecus geei*)
Golden monkey (*Pygathrix roxellana*)
Grizzled sureli (*Presbytis comata*)
Guereza (*Colobus guereza*)
Guinea forest red colobus (*Procolobus badius*)
Hanuman or Common langur (*Semnopithecus entellus*)
Olive colobus (*Procolobus verus*)
Proboscis monkey (*Nasalis larvatus*)
Satanic black colobus (*Colobus satanas*)
Silvered leaf monkey (*Semnopithecus cristatus*)

▲ This attractively colored Guinea forest red colobus is a 3-year-old male. The stump-like thumb typical of colobines is just visible. This species is found in Gambia, where the picture was taken, and other countries in West Africa.

eyes and nose are ringed with blue. Like the leaf monkeys, all the snub-nosed monkeys have a brow ridge.

THE SACRED APES

The leaf monkeys and langurs belong to the genus *Semnopithecus*, meaning "sacred ape". This is a fitting name for one of the most common of the Asian colobines, the Hanuman langur.

These langurs are found all through the southern foothills of the Himalayas, from Afghanistan to Tibet, and in India, Bangladesh and Sri Lanka.

In India the Hindus consider the Hanuman langur sacred, because they identify it with their monkey god Hanuman. The langurs often enter villages and towns. People feed them, especially on Tuesdays, which is the

god Hanuman's special day. The farmers, however, suffer badly from these "sacred apes," for they often feed on crops, and no one lifts a hand to stop them.

THE AFRICAN COLOBUS

Some of the most colorful colobine monkeys live in Africa, in the rain forests and savannah country on or near the equator. They are red colobus monkeys. Different species of these are to be found from Senegal in the west to Kenya in the east.

The coat of the red colobus is multi-colored, with the color varying from region to region. The paws, crown, back and tail-tip are often blackish-red. The brow may be white or orange, the cheeks and chest orange or yellowish-white. Various parts may also be tinged in shades of black, brown, gray or yellow.

▲The face of a Golden leaf monkey, showing its long, projecting crown and cheek hairs. The monkey is found only in Bhutan and neighboring Assam in north-east India. Like all colobines, it is threatened by a tourist demand for its pelt.

◄One of the most unmistakable of all monkeys, the Proboscis monkey. Only the male has the tongue-shaped nose.

A number of species of black colobus monkey also live in West, Central and East Africa. The Satanic black colobus is completely glossy black. Other species, such as the guereza, are black and white.

Many African colobus monkeys are hunted illegally for their beautiful coats. There is a tourist demand to use the coats for rugs or wall-hangings. In the 19th century several species were seriously threatened by the fur trade.

All colobus monkeys are strongly marked with one of the features that makes colobines different from other Old World monkeys. They hardly have a thumb at all – it is only a stump with a nail. This feature is present, but not so obviously, in the other colobines.

COURTSHIP

The females of most colobus and leaf-eating monkeys normally start the courtship process. This is unusual among animals. The females of other colobine species behave differently from the Proboscis monkey, however. For example, if the male Hanuman langur ignores the female's advances, she will hit him, pull his fur and even bite him.

TROOPS AND TAKE-OVERS

Among the colobines, China's rare Golden monkey forms the largest troops. Troops of more than 600 animals have sometimes been spotted. Outside China, the Hanuman langur may form troops of more than 100, and some of the red colobus species combine in groups of 50 or more. Other colobines live in small family groups of a breeding pair and one or two offspring.

The make-up of Hanuman monkey troops has been well studied. The main part of a troop consists of females, often with only one breeding male. The females remain in the same territory all their lives. Male offspring of the troop leave, or are driven away before they become adult.

The young males mostly join up with an all-male troop. These male groupings wander over large areas, keeping watch over several breeding troops. From time to time some of their number challenge the dominant males of the breeding troops and try to take them over. The females may help fight off the newcomers.

If and when a take-over does occur, the incoming male or males will often kill some of the infants in the troop. The killing of young after a male take-over also occurs in some leaf monkey species.

Biologists think that the incoming males may behave in this way to ensure that more of their own offspring will survive.

◀A Hanuman langur female and young. The infant may or may not be her own, because she will sometimes nurse the offspring of other females.

▲Two troops of colobines meet at the boundary between their territories. Fighting may take place but without any animal being injured.

▶A Hanuman langur troop. Most of the females will be related, and there will probably be just one breeding male.

GIBBONS

It is two hours after sunrise in the dense forest on one of the Mentawai Islands of Indonesia. A family of Kloss gibbons are feeding in the tree-tops. The male raises his head to whistle and sing, warning other males off his territory. He has been doing this, on and off, since before dawn. Soon it is the female's turn. Her song is even more musical, made up of long falling and rising notes and tuneful trills. During her final trills, she starts swinging through the branches, tearing off leaves. The rest of the family join in noisily.

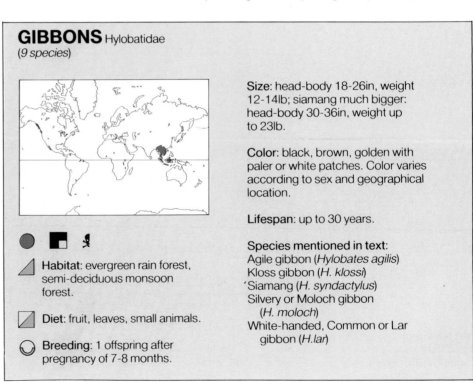

▼A White-handed gibbon in a typical gibbon position, hanging by its long arms. Both males and females look alike.

Gibbons are the only apes that live on the mainland of Asia. They are found from Bangladesh to Vietnam and along the Malay peninsula. They inhabit the islands of Sumatra and Borneo, along with the other Asian ape, the orang-utan.

Gibbons are much smaller and more lightly built than the orang-utan and the other great apes (chimpanzees and gorillas). They are often called the lesser apes. The two sexes are similar in size.

Like all the apes, they have no tail, and their arms are long. They often stand upright and can walk well on two legs, even along branches of trees. But their usual method of moving through the trees is by swinging, hand over hand, below the branches.

Most species of gibbons are about the same size. The exception is the siamang, which is nearly twice the size of the others. The siamang is completely black, as is the Kloss gibbon. More colorful is the Agile gibbon, which is light buff, with gold, reds and browns, and white cheeks and eye-

▲A siamang calls loudly from a bamboo thicket. The balloon-like sac on its throat makes the call resound and carry farther.

brows. There are some color differences between males and females of the same species.

LIFE-LONG MATE
The everyday life of the gibbon is centered around the family. Adult gibbons choose a mate for life, and produce young every 2 to 3 years. The

---

GIBBONS Hylobatidae
(*9 species*)

Habitat: evergreen rain forest, semi-deciduous monsoon forest.

Diet: fruit, leaves, small animals.

Breeding: 1 offspring after pregnancy of 7-8 months.

Size: head-body 18-26in, weight 12-14lb; siamang much bigger: head-body 30-36in, weight up to 23lb.

Color: black, brown, golden with paler or white patches. Color varies according to sex and geographical location.

Lifespan: up to 30 years.

Species mentioned in text:
Agile gibbon (*Hylobates agilis*)
Kloss gibbon (*H. klossi*)
Siamang (*H. syndactylus*)
Silvery or Moloch gibbon (*H. moloch*)
White-handed, Common or Lar gibbon (*H.lar*)

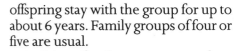

offspring stay with the group for up to about 6 years. Family groups of four or five are usual.

A gibbon family may not stay close together during the day. The animals go off on their own to feed. They come together from time to time to rest and groom, and many also sleep together.

FIERCE DEFENDERS
More than any other ape, gibbons are fierce defenders of their home territory. Both the male and female in a family group call loudly to warn others off. Their calls can be complicated and tuneful and are often described as songs. Each species of gibbon can be identified by the kind of song it sings. Male and female gibbons of most species perform a duet together, but the male and female Kloss gibbon sing solo.

◀A female Kloss gibbon launches herself into the air as she finishes her great call. Her baby, clinging to her belly, joins in.

▼The Silvery gibbon is named after its silvery-grey coat. Like several other gibbons, it is in danger of extinction because of logging operations.

CHIMPANZEES

Two groups of chimpanzees meet for the first time in days. They know one another, but there is tension in the air. A male from one group suddenly gets up and charges towards the other group. None of the opposing males wants to accept his challenge today.

▼ The face and ears of the adult Common chimpanzee are brownish-black. When it was young, they were pink.

The two species of chimpanzee both live in tropical Africa. The Common chimpanzee is found in West and Central Africa, north of the River Zaire. It inhabits thick forest and also more open savannah country.

The Pygmy chimpanzee, or bonobo, is found only in the rain forests of Zaire. Although called "Pygmy" it is not noticeably smaller than the Common species, but it is of slighter build. One main difference between the two species is in face color. The Common chimpanzee has a pink to brown face, the Pygmy chimpanzee an all-black one.

Both species spend much of the time on the ground. They sometimes walk upright on two feet, but they usually walk on all fours, using the knuckle not the palm of each hand.

STICKS AND STONES

Like the other apes, chimpanzees are mainly vegetarian, and they prefer to eat ripe fruit. But they also kill and eat animals such as monkeys, baboons, pigs and antelopes.

CHIMPANZEES
Pongidae (2 species)

Habitat: tropical rain forest, deciduous forest, mixed savannah.

Diet: fruit, leaves, flowers, seeds, some animals.

Breeding: 1 offspring after a pregnancy of 7½-8 months.

Size: head-body 28-34in, weight 66lb female; head-body 28-36in, weight 88lb male.

Color: coat black, graying with age.

Lifespan: up to 45 years.

Species mentioned in text:
Common chimpanzee (Pan troglodytes)
Pygmy chimpanzee or bonobo (P. paniscus)

Chimpanzees eat insects as well, such as caterpillars and ants. To reach ants inside a nest, they bring their stick-tools into action. They put the sticks into the nest and wait for the ants to crawl up. Chimpanzees also use sticks and sometimes stones to crack open fruit shells that are too hard to bite.

Once it was thought that only human beings used tools. But this use of sticks and stones shows how intelligent chimpanzees are. In fact, with gorillas, they are the second most intelligent creatures on Earth, after humans. Chimpanzees in captivity have learned to use some of the hand signals of the sign language of hearing impaired people.

GANG WARFARE

Every chimpanzee belongs to a large loose group or community of perhaps as many as 120 animals. Some live alone for much of the time, but most travel in small groups. There are mixed-sex family groups and also all-male groups of up to 12 that are a threat to other groups because they challenge the breeding males.

Displays of charging and stick throwing take place regularly in chimpanzee groups. They are usually performed by the strongest or most aggressive male as a challenge to other males. If they still accept him as boss, they bob up and down, panting and grunting. But if a male takes up the challenge, a noisy fight breaks out until one or the other runs away.

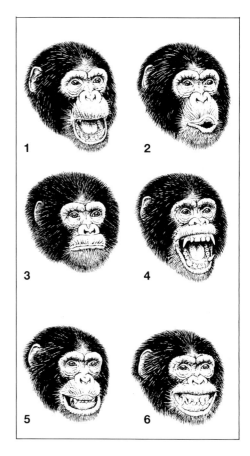

► The chimpanzee has one of the most expressive faces of all primates. (1) The relaxed play face. (2) The pout, used when begging for food. (3) The display face, which shows aggression. (4) The full open grin, showing fear or excitement. (5) The horizontal pout, showing surrender after being attacked. (6) The fear grin, displayed when approaching an animal of higher rank.

▼ Chimpanzees are threatened by destruction of their forest homes and by hunting for bushmeat.

ORANG-UTAN

A shaggy haired creature nearly as big as a man is swinging slowly through the branches of the tangled Borneo forest. It is the most colorful of the apes, the orang-utan. From time to time it stops to feed on insects or to raid a bird's nest. Then it continues on its way. After a few hours it arrives at its destination, a fig tree in full fruit. Drawing on its excellent memory and knowledge of the forest, the orang-utan knew exactly where to go. It also knew the exact time when the fruit would be ripe.

The orang-utan is a large, red, long-haired ape. It is found only on the islands of Sumatra and Borneo in South-east Asia. The name orang-utan means "man of the woods" in the Malay language. Those in Sumatra are thinner and have longer faces than the ones in Borneo. They also have a paler coat with longer hair.

The male adult orang-utan is very different in size and appearance from the female. He is almost twice as big and has large cheek flaps of fatty tissue. He also has a bag-like pouch hanging from the throat. When he makes his "long call," the air bag inflates and makes the call resound.

Adults' faces are bare and black, while the young have a pale muzzle and rings around the eyes.

▲A mother orang-utan and her infant. She will probably raise only three or four offspring during her 20 years of motherhood. At one time, mother orang-utans were shot to capture their young for use as pets.

▼A male orang-utan making his "long call". He breaks a branch off a tree and hurls it to the ground. Then he lets out a series of loud roars, bellows and groans.

ORANG-UTAN
Pongo pygmaeus

Diet: mainly fruit, also shoots, insects, eggs, other small animals.

Breeding: 1 offspring after pregnancy of 8½-9 months.

Size: head-body 31in, weight 88-110lb, height 46in female; head-body 39in, weight 132-198lb, height 55in male.

Color: orange-red, bright in the young, dark in adults. Face bare and black but pinkish on muzzle of young animals.

Lifespan: up to 35 years.

Habitat: tropical rain forest.

HIGH INTELLIGENCE

The orang-utan's ability to remember the location and fruiting times of individual trees in thick forest is one example of its high intelligence. In captivity these animals prove to be quick learners and good companions of humans.

In the wild, orang-utans have a very simple life-style and spend most of their time on their own. Males and females come together to mate, but when the female gets pregnant she goes off on her own to give birth and raise the baby. A male will usually maintain a relationship with several females in his territory. He makes his calls both to attract more females and to warn off other males.

Males usually try to avoid one another, and their long calls help them to keep their distance. If two males do meet, however, they kick up a rumpus, charging about, shaking and breaking off branches, calling and sometimes (though not often) fighting each other.

THREAT OF EXTINCTION

Orang-utans are shy, gentle creatures, and local people treat them with respect. Yet like the other "gentle apes", the gorillas, orang-utans are in danger of becoming extinct.

Once they were threatened by people who trapped them for export to zoos. But this has now mostly been stopped. Today, the greatest threat comes from the destruction of their rain forest habitat due to logging.

To improve the orang-utan's chances of survival, conservationists have set up nature reserves and national parks in Borneo and Sumatra and elsewhere in Indonesia and Malaysia.

►The orang-utan spends most of its time in the trees. It swings on its long arms and takes hold of branches with its hook-like hands and feet.

GORILLA

Twice as big as the females around him, a mature "silverback" gorilla sits in the forest chewing a mouthful of leaves. Hearing a crashing in the undergrowth, he jumps to his feet. Advancing on him threateningly is a much younger "silverback". When they see each other they roar loudly, beating on their chests with both hands. The younger gorilla knows he's outclassed.

The gorilla is the largest of all the primates. Mature female gorillas are heavier than most human males. Mature males – known as silverbacks because of their silvery-white saddle – are twice as heavy again. Along with chimpanzees, gorillas are the most intelligent animals next to human beings.

Gorillas live in groups, usually of 5 to 10 animals. Each group is made up of one adult male and a number of females and their young.

GENTLE GIANTS

For most of the time, gorillas lead a peaceful life. They spend the greater part of the day feeding, since they need to eat a lot of food, mainly leaves, to maintain their huge bulk. In between they rest to allow plenty of time for digestion. At night they make platform nests of twigs and branches either on the ground or in low trees.

Gorillas spend more time on the ground than up in the trees. They occasionally walk on two feet, but for the most part they walk on all fours. Like chimpanzees, they walk on the soles of their feet and on the knuckles of the hands. Only young gorillas spend much time up in the trees.

Threat displays between rival gorillas can sometimes end in ferocious fights. The worst clashes occur between the silverback with a well-

▲An Eastern lowland gorilla eating a plant stem. Gorillas have large teeth and powerful jaws to crunch the huge amounts of leaves and stems they eat. Their jaws are worked by large muscles attached to a bony crest on the skull.

established "harem" and the lone young silverback. The loner tries to steal some or all of the other's females and set up a harem of his own.

In the wild gorillas do not feel threatened by human beings, and will allow them to approach quite close. But human visitors must stay quiet,

▼A silverback Mountain gorilla, leader of his "harem" of smaller females.

GORILLA *Gorilla gorilla*

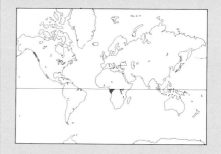

⬤ ◧ ☠

◺ Habitat: tropical forest.

▢ Diet: mostly leaves, some fruit.

◯ Breeding: 1 offspring after pregnancy of 8-9 months.

Size: height up to 72in male, 60in female; weight up to 395lb male, 198lb female.

Color: coat black to brownish-gray.

Lifespan: up to 40 years.

Races mentioned in text:
Eastern lowland gorilla (*Gorilla gorilla graueri*)
Mountain gorilla (*G. g. beringei*)
Western lowland gorilla (*G. g. gorilla*)

◀▼A gorilla group rests at midday after a morning's feeding. They gather, with the females and babies closest, around the silverback, the only mature male. One of the females grooms him. Older infants play together. Females with no offspring and maturing males stay on the edge of the group.

must sit or squat and above all must not stare. Staring is considered very rude in gorilla society.

SHRINKING HABITAT

The gorilla is a single species, but there are three separate races or subspecies. All are under threat, like so many other animals, from the destruction of their forest habitat. Hunting and trapping the young for sale to zoos are also reducing their numbers.

Most under threat is the Mountain gorilla, of which only about three or four hundred remain. The Mountain gorilla is found in Zaire, Rwanda and Uganda at altitudes between about 5,000ft and 12,500ft. A few thousand Eastern lowland gorilla remain, living in eastern Zaire. The Mountain and Eastern lowland gorillas have black coats. The male's silvery-white saddle is only on the back.

The Western lowland gorilla has a brownish tinge to its coat, and the male's saddle extends to the rump and thighs. It is found in Central West Africa from Cameroon to the Congo. About half of the total population (around 9,000) are concentrated in Gabon.

▶A Mountain gorilla nursing her baby. She will feed it for up to 3 years and probably have another baby a year or so later. Gorillas make caring and affectionate mothers, but their babies do not always survive. In her lifetime a female gorilla may successfully raise only three or four offspring.

TENRECS

The female Common tenrec has been pregnant for almost two months, and her time is due. She settles in her cosy nest and soon feels the first pangs of birth. Then the babies start appearing – one, two, three, four, five... more than 20 babies emerge from her body. Although all but helpless, they manage to find the mother's teats and take their first feed of milk. None goes without, for there are more than enough teats to go round.

The Common tenrec is the largest member of the major animal group of insectivores or insect-eaters. Like most other insectivores it has a long narrow snout covered with sensitive whiskers. Its eyesight is not good, and it relies more on its excellent hearing and sense of smell to detect its prey. Its brown coat is coarse and has spines along the back.

When fighting, it opens its mouth in a wide gape, exposing long canine teeth. It slashes sideways at its opponent, then bucks with its head. This brings into action the stiff spines on its neck. Similar behavior is found in other tenrecs.

MADAGASCAN HEDGEHOGS

The Common tenrec is found in large numbers on the island of Madagascar in the Indian Ocean. It is one of 31 tenrec species on the island. One of its close relatives is the Streaked tenrec, so called because of the white stripes on its blackish coat.

The Streaked tenrec looks much like the young Common tenrec. Both practise stridulation. This means they rub together stiff spines on their back to make a noise. The young do this when alarmed or just to communicate with one another and with their mother when foraging.

TENRECS Tenrecidae
(*34 species*)

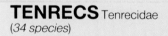

Habitat: wide range, from semi-arid to rain forest, also mountains, rivers and human settlements.

Diet: worms, insects, birds, crustaceans, some fruit.

Breeding: 2-32 offspring after pregnancy of 7-9 weeks.

Size: smallest (*Microgale parvula*): head-tail 3½in, weight ⅕ ounce; largest (Common tenrec): head-tail 16in, weight 3¼lb.

Color: brown to gray, sometimes streaked.

Lifespan: up to 6 years.

Species mentioned in text:
Aquatic tenrec (*Limnogale mergulus*)
Common tenrec (*Tenrec ecaudatus*)
Greater hedgehog tenrec (*Setifer setosus*)
Large-eared tenrec (*Geogale aurita*)
Lesser hedgehog tenrec (*Echinops telfairi*)
Long-tailed tenrec (*Microgale melanorrachis*)
Rice tenrec (*Oryzoryctes tetradactylus*)
Streaked tenrec (*Hemicentetes nigriceps*)

Two other members of the tenrec family have stiff spines all over their body except the underside, just like a hedgehog. They are called the Greater and Lesser hedgehog tenrecs. The latter spends some of its time in the trees. Both roll themselves into a protective ball when threatened.

Quite different in appearance are the soft-furred Long-tailed and Large-eared tenrecs, which look rather like shrews. Some are able to climb trees, while others sometimes burrow in the ground. The Rice tenrec is a more persistent burrower, while the Aquatic tenrec is a water creature, whose webbed feet make it a powerful swimmer.

RECORD BREEDERS

The female Common tenrec has been known to produce as many as 32 young in one litter. This is a record among mammals. Like most tenrecs, the Common tenrec is usually nocturnal. But nursing mothers with large families have to forage in the daylight hours as well. Their young have a striped coat, which gives them good camouflage in the relatively dangerous daylight hours. The young tenrecs become more nocturnal as they acquire the darker adult coat.

▼When cornered, the Lesser hedgehog tenrec curls itself up into a ball. Its prickly spines keep off most predators.

◄Species of tenrec Tenrecs vary widely in appearance, from the hedgehog-like Lesser hedgehog tenrec (1) to the shrew-like Long-tailed tenrec (2). Largest is the Common tenrec (3), here displaying its powerful jaws. The Streaked tenrec (4), is easy to recognize. The Rice tenrec (5) is also shrew-like. The aquatic Giant otter shrew (*Potamogale velox*) (6) and Ruwenzori least otter shrew (*Micropotamogale ruwenzorii*) (7) live in streams, lakes and swamps in tropical Africa.

SOLENODONS

A rather large rat-like creature is feeding among the leaf litter and stones on the forest floor. It lifts up its head, showing an unusually long snout. The animal is a solenodon. After sniffing the air, it goes back to using its snout to probe under stones and into crevices. Soon its labours are rewarded, as a beetle scurries out from shelter. The solenodon lunges and pins down the beetle with its snout, then scoops up its prey into its mouth. The beetle soon stops struggling, poisoned by the solenodon's saliva.

Solenodons live in the forests of Hispaniola (Haiti and the Dominican Republic) and Cuba. The solenodons on Hispaniola belong to a different species from those on Cuba, but they are closely related. Both species have the same sized body, but the Cuban species has a shorter tail. They also have a different colored coat. The Cuban solenodon's coat is dark gray, except for a pale head and belly. The Hispaniola solenodon has a much coarser coat, which is gray-brown on the back and yellowish on the flanks; the forehead is black.

SOUND EFFECTS

Solenodons are active at night. Their eyes are small, and their eyesight is poor. They learn about their surroundings mainly with their other senses. They have a good sense of touch through the long sensitive whiskers on the snout. Their senses of smell and hearing are also well developed, helping them locate their prey in the dark.

An interesting sound solenodons make is a high-pitched clicking noise.

▼When hunting, solenodons use their long sharp claws to upturn stones and tear off bark in search of prey.

SOLENODONS
Solenodontidae (*2 species*)

● ■ ☠

◢ Habitat: remote forest.

■ Diet: millipedes, beetles and other insects, worms and termites, sometimes birds.

◗ Breeding: 1 or 2 offspring; length of pregnancy unknown.

Size: head-body 11-13in; tail 9-10in; weight 1½-2lb.

Color: Hispaniola species mainly brownish-gray; Cuban species dark gray.

Lifespan: unknown.

Species mentioned in text:
Cuban solenodon (*Solenodon cubanus*)
Hispaniola solenodon (*S. paradoxurus*)

◄The Hispaniola solenodon is being killed and driven out of Haiti mainly by dogs. It has some protection in the neighboring Dominican Republic.

They probably use this as a kind of sonar to find their way around. They give out the clicks and listen for echoes, which will tell them if there is something in their path. Solenodons also twitter, chirp and puff as they move around.

DWINDLING NUMBERS

Before the Europeans arrived in the Caribbean region, solenodons were found on several other islands besides Hispaniola and Cuba. But the animals the Europeans introduced – cats, dogs and mongooses – preyed on the solenodons and killed them off. The same predators have today reduced the remaining solenodon populations to danger levels. On Hispaniola and Cuba they are found now only in the remotest regions.

Another reason why solenodons have become scarce is that they have a much lower rate of reproduction than their predators. Only one or two offspring are produced at a time. The young rely on their mother for several months, much longer than with other insectivores.

During the first few weeks of their life, young solenodons are carried around by their mother in a unique way. They cling to her long teats. Later they follow her on foot around the nest burrow. They get a taste of grown-up food by licking the mother's mouth as she feeds.

◄One of the most unusual features of the solenodon is its long, tapering and whiskered snout. The snout is made of cartilage, not bone, and extends beyond the jaw. In the Hispaniola species it is joined to the skull by a ball-and-socket joint, which makes it very flexible.

HEDGEHOGS

Crawling into the nest it has made under a pile of logs, the European hedgehog settles down and falls asleep. It is late fall, and the temperature is dropping fast. Next day, the hedgehog is still asleep, and the next, and the next. Its hibernation may last until the spring. During this time it appears dead, but its heart is still beating – albeit slowly.

Hedgehogs are among the most familiar small mammals there are. Their stiff spines give them good protection against most predators, so they are not too worried about being caught out in the open. When they are attacked, they roll themselves into a prickly ball, shielding the soft underparts.

Hedgehogs have the long snouts that most insectivores have. But although insects form a major part of their diet, they will also feed on other small creatures and have even been known to attack quite large birds.

Hedgehogs are adaptable animals, which is why they are found so widely throughout Europe, Asia and Africa. Most is known about the European hedgehog.

HEDGEHOGS Erinaceidae (*about 17 species*)

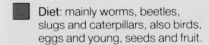

Habitat: from desert and steppe to farmland and forest.

Diet: mainly worms, beetles, slugs and caterpillars, also birds, eggs and young, seeds and fruit.

Breeding: up to 10 offspring after pregnancy of 5-7 weeks (European hedgehog).

Size: smallest (Lesser moonrat): head-body 4in, tail ½in, weight 1½ ounces; largest (Greater moonrat): head-body 18in, tail 8in, weight 3lb.

Color: coat brown, paler on underside; Greater moonrat black and white.

Lifespan: up to 8 years.

Species mentioned in text:
European hedgehog (*Erinaceus europaeus*)
Greater moonrat (*Echinosorex gymnurus*)
Lesser moonrat (*Hylomys suillus*)
Mindanao moonrat (*Podogymnura truei*)

HIBERNATION

The European hedgehog often hibernates, or sleeps, through the cold of winter, because during this time food becomes scarce. Its relatives in other regions with a cold season also hibernate. Hedgehogs that live in Africa's tropical regions, however, have no need to hibernate, because there is a plentiful supply of food for them all year round.

In desert regions hedgehogs may sleep through the driest and hottest periods of the year. This summer sleep is known as estivation.

To prepare for hibernation, the hedgehog eats well in the summer, almost doubling in weight. It builds up a thick layer of fat beneath the skin, which sustains it during its long sleep. The layer also acts as insulation to keep it warm when the temperature outside drops to freezing and below.

The hedgehog's own temperature may fall until it is only a few degrees higher than its surroundings. In this state it is barely alive. It takes only a few shallow breaths every minute, and its heart beat is so slow and feeble that it can scarcely be felt.

STRANGE HABITS

There are many folk tales about the hedgehog's behavior. They are supposed, for example, to carry fruit on their spines and to suck the milk from

◄This all-white Greater moonrat is unusual. Most members of the species are mainly black, with white just on the head and shoulders.

cows, but this seems most unlikely. They certainly have one or two strange habits, though, such as self-anointing. They produce lots of foamy spit and then flick it over their spines with the tongue. Nobody knows why they do this. It could be a way of attracting the opposite sex, of cleaning the spines or of ridding the skin of parasites.

The European hedgehog is equally at home in the town garden or in the countryside. Many people treat hedgehogs as pets, leaving out bread, milk and other food for them. People often think that "their" hedgehog lives somewhere near by, but this is not necessarily true. During their search for food, which for many species is at night, hedgehogs may travel as far as 1 mile from their nest site, sometimes

▶**Species of hedgehogs and moonrats**
The Desert hedgehog (*Paraechinus aethiopicus*) **(1)** often digs and lives in burrows. The Shrew hedgehog (*Neotetracus sinensis*) **(2)** is much smaller and has a softer hairy coat, not spines. It is a kind of moonrat and has a long tail, as do the Mindanao moonrat **(3)** and the Greater moonrat **(4)**.

farther. Also, they change their nest site frequently.

Hedgehogs are mostly solitary animals. The male and female come together only to mate. During courtship the female keeps her spines erect if she does not want to mate and may butt the male with them. She flattens her back spines when she is ready to mate. The male gets on her back, but

has to hold on to the spines on the female's shoulders to prevent himself slipping off.

After mating, the male departs, showing no interest in bringing up the young. The female prepares a nest in which to give birth. She uses dry leaves, grasses and moss to create a warm dry bed, usually beneath a bush but sometimes underground.

SPINELESS YOUNG

When the young are born they are naked and blind. Their spines lie beneath the skin to prevent them damaging the mother during birth. They are embedded in a watery fluid, but this soon disappears, and 150 or so white spines soon force their way through the skin. Within 2 days more dark spines start to grow out.

The young stay with the mother until they are about 7 weeks old (European hedgehog), then they leave or are driven away by her. The European hedgehog may breed twice a year if the food supply is good. The hedgehogs that live in the tropics breed all through the year.

THE HAIRY HEDGEHOGS

Related to the familiar spiny hedgehogs are a group of animals that look more like rats. They have a coat of hair, not spines, and are often called hairy hedgehogs. These are the moonrats or gymnures, animals native to China and South-east Asia.

Like their spiny relatives, moonrats have a well developed sense of smell. They are also very smelly themselves. They have scent glands that give out what is to humans a quite unpleasant scent, sometimes described as being like rotting garlic.

The Greater moonrat is the largest member of the hedgehog family, being about the size of a cottontail rabbit. Not a great deal is known about the habits of moonrats because they are very shy creatures, unlike hedgehogs. Some species are under threat of extinction because of the destruction of their jungle habitat. The Mindanao moonrat of the Philippines seems in the most danger.

▶A family of European hedgehogs goes foraging in the grass. From the age of about 3 weeks the young leave the nest and follow their mother. There may be up to seven young in the litter.

SHREWS

Ever since it was born a year ago, the little shrew scurrying through the leaf litter has been on the go. Every hour or so it has had to eat to give it the energy for its very active way of life. Now it is in trouble. Its one set of teeth has worn down almost to nothing. It can hardly chew even the softest prey. Suddenly the scurrying stops. The shrew drops in its tracks, dead of starvation.

As a zoological family, shrews are by far the most successful of the insect-eaters or insectivores. Out of the 345 species of insectivores, 246 are shrews. They are found throughout North and Central America, Europe, Asia and most of Africa. Among the 246 species there are over 20 distinct types or genera.

Some species can be found widely, others only in certain areas. The Eurasian water shrew is found throughout Northern Europe and Asia; the Sri Lanka shrew, only on that island.

SQUEALS AND ECHOES
All the shrews are broadly similar in appearance. They have a greyish or brownish soft-furred body and a tail.

They look at first sight like a mouse, but have a much more pointed nose. Their eyes and ears are small and often nearly hidden by their fur. Their sight is poor, but their hearing is good.

Shrews spend most of their time alone. They squeal and twitter when they come across other shrews wandering into their territory. Some use sonar (echo-location) to find their

way around. They make high-pitched clicking noises and then listen for the echoes from obstacles in their path.

▼ This shrew seems to have bitten off more than it can chew. But large juicy earthworms are one of its favourite foods, and it is quite able to cope. Shrews need to eat every few hours to stay alive. They often eat more than their own body weight every day.

SHREWS Soricidae
(246 species)

○ ▢

■ **Habitat:** wide range, from desert and grassland to forest.

■ **Diet:** worms, insects and other small animals, sometimes seeds and nuts.

◎ **Breeding:** up to 10 offspring after pregnancy of 2-3 weeks.

Size: smallest (Pygmy white-toothed shrew): head-tail 3.5cm, weight 2g; largest (African forest shrew): head-tail 29cm, weight 35g.

Colour: brown, grey.

Lifespan: up to 18 months.

Species mentioned in text:
American short-tailed shrew (*Blarina brevicauda*)
Armoured shrew (*Scutisorex somereni*)
European or Eurasian water shrew (*Neomys fodiens*)
Pygmy white-toothed shrew (*Suncus etruscus*)
Sri Lanka shrew (*Podihik kura*)
Tibetan water shrew (*Nectogale elegans*)

▲The Pygmy white-toothed shrew is the smallest land mammal on Earth, measuring 1½in long and weighing ¹⁄₁₅ ounce. It lives in African scrub and savannah.

▼A mother gives young shrews a guided tour of their territory. Each animal grips the rump of the one in front, so none gets lost on the journey.

The water shrews are well adapted for their life in the water. Stiff hairs fringe their feet and tail, helping them swim better. The Tibetan water shrew even has webbed feet.

The Armored shrew of Central Africa boasts the most unusual feature among shrews. It has enlarged vertebrae on the backbone, which have tough spines attached. It is said that they can support the weight of a man lying on top.

FIERCE FIGHTERS

The only time one usually sees shrews is when they have dropped dead in the open. When they are alive, shrews are secretive creatures that stay out of sight among the leaf litter if there is a disturbance near by. Dead shrews often remain uneaten for a long time because of the evil-smelling scent that comes from glands in their body. Many predators leave them alone for the same reason.

Some shrews have not only glands like this, but also poisonous saliva. The American short-tailed shrew and some water shrews have venomous bites that help them attack quite large animals such as mice, frogs and fish. For their size, shrews are among the fiercest animals in the world.

Several species of shrew are known to nibble and lick their rectum as soon as their intestines are free of droppings. They probably do this to obtain minerals and vitamins from partly digested food and droplets of fat in the rectum.

◄A European water shrew dives into a stream for food. It will feed on small fish, crustaceans and small frogs. Note the silvery air bubbles clinging to its fur.

49

Three feet under the ground a female mole suckles her five young. Soon she leaves them in the nest chamber and climbs up the sloping passage to her feeding grounds. There the tunnels are just beneath the surface and act as pitfall traps for the insect larvae and worms on which she feeds. But her catch is not enough to satisfy her, so she digs a fresh tunnel in search of richer pickings before returning to feed her family.

▼A European mole emerges from the middle of a molehill. Its powerful fore-limbs are spade-like and tipped with strong claws.

MOLES AND DESMANS Talpidae and Chrysochloridae (*47 species*)

○ ■ 🕱

■ Habitat: moles: mostly sub-terranean, but from desert (golden moles) to grassland and forest; desmans: lakes and rivers.

■ Diet: moles: worms, insects, slugs, snails, spiders, lizards; desmans: water insects, crustaceans, and sometimes fish.

◎ Breeding: up to 7 offspring after pregnancy of 1 month (European mole).

Size: smallest (shrew moles): head-body 1in, tail 1in, weight 1/3 ounce; largest (Russian desman) head-body 8½in, tail 8in, weight 19 ounces.

Color: gray, brownish-black; golden moles: reddish-brown, golden.

Lifespan: 3-5 years.

Species mentioned in text:
European mole (*Talpa europaea*)
Grant's desert golden mole (*Eremitalpa granti*)
Pyrenean desman (*Galemys pyrenaicus*)
Russian desman (*Desmana moschata*)
Star-nosed mole (*Condylura cristata*)

The European mole spends most of its life underground, and its body is well suited to this life. It has a rounded body covered in thick velvety fur. Its most noticeable feature is its huge forelimbs, with which it burrows through the soil.

When digging, the mole first braces itself with its back legs. Then it moves its forelimbs forwards into the soil and then sideways and backwards in almost a swimming motion. From time to time it digs a shaft upwards and pushes the loose soil it has excavated to the surface. This results in the familiar molehill.

BLIND AS A MOLE
Moles are very active creatures and eat frequently. The European mole spends about four hours eating followed by four hours resting, day and night for much of the year. On occasions it may burrow nearly 330ft a day if food is scarce.

Moles have very small eyes, almost hidden by their fur, and they are all but blind. For a life underground this does not matter. Their main sense is that of touch. They have sensitive whiskers on the snout and also sensitive little bumps there.

Moles' sense of smell is also good.

They smear their burrows with scent to warn other moles away. Moles have some sense of hearing, although their ears are little more than fur-covered holes.

RIVER AND DESERT MOLES
Some moles are excellent swimmers. The Star-nosed mole lives in tunnels in river banks and swims a lot. The desmans, the largest of the moles, are even more aquatic. They have webbed feet, a flat paddle-like tail and a trumpet-like snout which they use like a snorkel. One species lives in Russia, the other in the Pyrenees mountains. Both are now scarce.

None of the moles mentioned so far are found in Africa. On that continent the moles are represented by 18 species of the almost tailless golden moles. The name "golden" is taken from the grayish-yellow coat of Grant's desert golden mole, which lives in South-west Africa.

▲The Star-nosed mole (1) is equally at home on land or in the water. The Pyrenean desman (2) spends nearly all its time in the water.

▼This Star-nosed mole has caught a worm. It has 22 fleshy tentacles on the end of its nose.

ANTEATERS

The Giant anteater lumbering across the dried grassland is heading for one of the anthills dotting the landscape. It can't see the mound because of its poor eyesight, but it can smell the ants. When it reaches the earthy mound, it slashes a hole in it and thrusts in its snout. Flicking its long tongue in and out, it begins to feed on the startled insects.

The Giant anteater is the largest of the anteaters of South and Central America. They all feed almost entirely on ants and termites. They have a tube-like snout and a long narrow tongue covered with a sticky saliva. The Giant anteater can flick its tongue a distance of some 24in up to 150 times a minute.

The ants stick to the saliva on the tongue and are taken into the mouth.

The mouth opening is surprisingly small – not much bigger across than a pencil. Anteaters have no teeth. They lightly chew their prey using little hard lumps on the roof of the mouth and on the cheeks.

▼ The Southern tamandua's gold, brown and white patterned coat gives it good camouflage in the scrubland and forest where it lives.

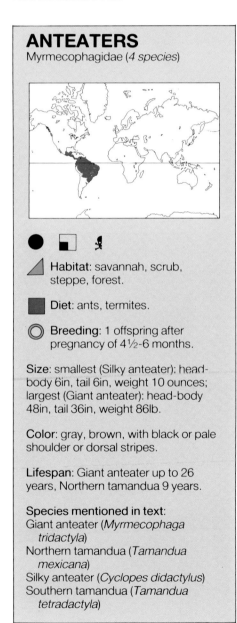

ANTEATERS
Myrmecophagidae (*4 species*)

● ■ ♞

◤ **Habitat:** savannah, scrub, steppe, forest.

■ **Diet:** ants, termites.

◎ **Breeding:** 1 offspring after pregnancy of 4½-6 months.

Size: smallest (Silky anteater): head-body 6in, tail 6in, weight 10 ounces; largest (Giant anteater): head-body 48in, tail 36in, weight 86lb.

Color: gray, brown, with black or pale shoulder or dorsal stripes.

Lifespan: Giant anteater up to 26 years, Northern tamandua 9 years.

Species mentioned in text:
Giant anteater (*Myrmecophaga tridactyla*)
Northern tamandua (*Tamandua mexicana*)
Silky anteater (*Cyclopes didactylus*)
Southern tamandua (*Tamandua tetradactyla*)

"STINKERS OF THE FOREST"

The Giant anteater is usually active during the day, but it spends part of the time asleep. The other species of anteaters are mainly nocturnal and spend much of their time in the trees. Unlike their giant relative, they have a prehensile tail which helps them grip the branches when they climb.

The two species of tamandua, the Northern and Southern, are only about half the size of the Giant anteater. They have a striped coat, which gives them their alternative name: Collared anteater. They have also been given the nickname "stinker of the forest" because of the unpleasant smell they sometimes give off.

The much smaller Silky anteater hardly ever comes down from the trees. Its snout is much shorter than that of the other species. It is often called the two-toed anteater because three of its five fingers do not show. It has short silky fur.

POWERFUL CLAWS

All the anteaters have large sharp and powerful curved claws on their fore-feet; tamanduas have three, the Giant and Silky anteaters two. They use these claws to open up anthills and termite mounds and also to defend themselves.

When alarmed or attacked, ant-eaters rise up on their hind legs, using their tail as a prop to steady them-selves. As their attacker gets closer, they slash at it with their claws, which on the Giant anteater are up to 4in long. Another powerful weapon is a crushing bear-hug, delivered with their strong forelimbs.

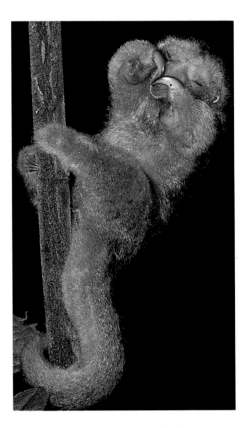

▶ On the defensive, a Silky anteater covers its face with its claws, while clinging to a branch with feet and tail.

▼ A young Giant anteater rides piggy-back on its mother. Both are identical in color, making the young one rather difficult to see.

ARMADILLOS

It is late evening in the heart of Florida's swampland. Coming to a stream, an armadillo stops as if wondering what to do next. Then it steps into the water and disappears beneath the surface. Holding its breath, it walks along the bottom of the stream. More than five minutes pass before it reappears on the other side and continues on its way. Had the stream been any wider, the armadillo would have swum across, swallowing air to make it float better.

▼ Southern three-banded armadillo (1), pichi (2) and Lesser fairy armadillo (3).

There are 20 species of armadillo in the Americas, ranging from Oklahoma in the north to Argentina in the south. Most widespread is the Common long-nosed armadillo, also called the Nine-banded armadillo, found in the United States.

The Spanish word *armadillo* means "little armored one". This is a very good description of an animal that is covered with a number of hard bony plates, called scutes. Broad shield-like plates usually cover the shoulders and rear of the body. In the middle are a varying number of circular bands, which flex as the animal moves. The head, tail and limbs are also protected by armor. The underside of the body is covered only by hairy skin, but an attacker is rarely able to reach this weak-spot.

GREAT DIGGERS

Armadillos have short but powerful limbs, tipped with strong claws. The animals use their claws when digging for insect prey or making a burrow to sleep in. The Common long-nosed armadillo is an especially efficient digger. When it smells insects or other small prey in the soil, it digs frantically for them, keeping its long nose pressed to the ground. It holds its breath while digging, to stop itself inhaling the dirt.

The burrows armadillos dig can be as much as 6½ft underground and have two or more entrances. They contain one or two nest chambers lined with grass and other plant material. An armadillo may dig 10 or more such burrows, which it uses on different days in no fixed pattern.

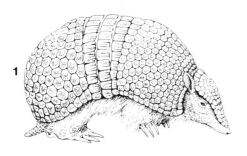

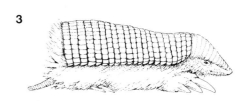

1

2

3

ARMADILLOS
Dasypodidae (*20 species*)

● ◨ ☠

◨ Habitat: wide range, desert, savannah, scrub, forest.

■ Diet: insects, especially ants and termites, other small animals.

◎ Breeding: number of offspring varies, 1 (fairy and three-banded armadillos), 4 (most long-nosed armadillos), 12 (Southern lesser long-nosed armadillo). Pregnancy usually 60-65 days.

Size: smallest (Lesser fairy armadillo): head-body 5in, tail 1in, weight 3 ounces; largest (Giant armadillo): head-body 40in, tail 18in, weight 132lb.

Color: pinkish or yellowish dark brown armor, pale or dark brown hairs between plates and on underside skin.

Lifespan: up to 15 years.

Species mentioned in text:
Brazilian lesser long-nosed armadillo (*Dasypus septemcinctus*)
Common long-nosed or Nine-banded armadillo (*D. novemcinctus*)
Giant armadillo (*Priodontes maximus*)
Larger hairy armadillo (*Chaetophractus villosus*)
Lesser fairy or Pink fairy armadillo (*Chlamyphorus truncatus*)
Pichi (*Zaedyus pichiy*)
Southern lesser long-nosed armadillo (*Dasypus hybridus*)
Southern three-banded armadillo (*Tolypeutes matacus*)

▲ Like all the armadillos, this Larger hairy armadillo has large strong claws for digging. It can take a variety of foods, including maggots from inside the rotting carcasses of other animals. It sometimes burrows deep within a carcass.

▼ The Brazilian lesser long-nosed armadillo digs a shallow burrow to rest in during the day. Like all the other armadillos it is classed as an *Edentate* (without teeth) but has a set of up to 100 primitive teeth.

GIANTS AND FAIRIES

Armadillos spend most of their lives alone. They mark their home territory with their urine and droppings and with a yellowish smelly liquid given off by glands at their rear. When armadillos do cross into each other's territory, they may fight, with much kicking, chasing, and squealing.

When an armadillo is being hunted by a predator it may dig itself out of trouble and disappear beneath the soil. Or it may simply crouch low on the ground so that only its armor shows. Three-banded armadillos can roll themselves completely into a ball, safe even from jaguars.

There is a great difference in size among armadillos. Largest of the family is the increasingly rare Giant armadillo which measures up to 60in from head to tail. The smallest is only one-tenth this length. This is the almost shrimp-like Lesser fairy armadillo, also called the Pink fairy armadillo, which has a dense coat of white hair on its sides and underparts. The Lesser fairy spends much of its life tunnelling underground – something its giant relative could never do.

PANGOLINS

Two lion cubs are padding through the grassy savannah. They spy a curious scaly creature shuffling along in front of them. It is a pangolin. When it hears them coming, the pangolin rears up on its hind legs and starts running. Its long tail acts like a prop to help it stay upright.

The pangolin has been described as a "walking pine-cone" or "walking artichoke" because of the overlapping scales that cover all except the underside of its body. Its favorite foods are ants and termites. It has no teeth, but grinds up the insects with the hard lining of its stomach.

THE TREE PANGOLINS

Pangolins live in Africa and Asia, in the dense tropical rain forests and in the more open grasslands. They are active mainly at night. The African rain forests are home for two species of pangolin which stay mainly in the trees.

The Long-tailed pangolin spends most of its time high up in the forest canopy. It feeds from nests of ants and termites hanging from the branches. The larger Small-scaled tree pangolin keeps more to the lower branches of the trees. Both species have a long

▲A Small-scaled tree pangolin. Pangolins' overlapping brown scales give them an unusual appearance.

▼A Cape pangolin drinks at the Etosha National Park in Namibia. Outside such parks, people hunt pangolins widely for their meat and scales.

PANGOLINS Manidae
(7 species)

● ◼ ☠

◼ Habitat: wide range, from forest to open savannah.

◼ Diet: mainly ants and termites.

◎ Breeding: usually 1 offspring after pregnancy of 4½ months (Cape pangolin).

Size: smallest (Long-tailed pangolin): head-body 12in, tail 22in, weight 2lb; largest (Giant pangolin): head-body 34in, tail 32in, weight 72lb.

Color: light yellowish-brown, through olive to dark brown, with white to brown undersurface hair.

Lifespan: up to 13 years in captivity (Indian pangolin).

Species mentioned in text:
Cape pangolin (*Manis temmincki*)
Chinese pangolin (*M. pentadactyla*)
Giant pangolin (*M. gigantea*)
Indian pangolin (*M. crassicaudata*)
Long-tailed pangolin (*M. tetradactyla*)
Malayan pangolin (*M. javanica*)
Small-scaled tree pangolin
   (*M. tricuspis*)

prehensile tail, with a sensitive bare patch at the tip which helps them climb.

THE GROUND DWELLERS
The two other African species of pangolin are bigger still and they are ground dwellers. They are the Cape pangolin and Giant pangolin. They use their powerful foreclaws to destroy termite mounds and ants' nests and then take up the insects with their long tongues, which are covered with sticky saliva. The Giant pangolin has an unusually long tongue – up to 28in. It may eat as many as 200,000 ants in a single night.

The pangolins of Asia are natives of central and southern India, southern China and as far south as Malaysia and the Indonesian islands. The name pangolin comes from a Malay word meaning "rolling over," which is what pangolins do when threatened.

The three species of Asian pangolin, that is the Indian, Chinese and Malayan pangolins, are similar in appearance. They live mainly on the ground but can climb skilfully. They differ from their African relatives in having hair at the base of their scales.

SMELLS AND SPRAYS
Pangolins live for most of the time alone. Their main sense is that of smell. They mark their territory with their droppings and urine and with a foul-smelling liquid from glands at the base of their tail.

They will also spray this liquid and urine in the face of animals that attack them. The larger ones may also lash out at their attackers with their tail. But pangolins usually rely on curling into a ball for defense. The overlapping scales on its body form a shield that is impregnable to all but the larger cats and hyenas.

▲A Small-scaled tree pangolin hangs by its tail from a branch. The tail has a short bare patch at the tip to make gripping easier.

57

BATS

It is the darkest time of night. The herd of cattle have settled down to rest. Out of the darkness comes the faint flutter of tiny wings. A vampire bat is swooping in for its night-time feed. It settles lightly on the shoulders of a young calf. With a swift movement of its tiny head, it slits the animal's hide with razor-sharp teeth. The calf doesn't feel a thing. The vampire's tongue then sets to work, flicking in and out of the wound as the blood starts oozing out.

▶The Mexican Long-nosed bat uses its enormous tongue to reach the nectar in desert flowering plants such as cactus.

BATS Megachiroptera and Microchiroptera (*up to 1,000 species*)

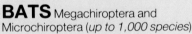

Habitat: general.

Diet: most bats: insects; vampire bats: blood; flying foxes: fruit; also eaten by different species: nectar, spiders, frogs, lizards, fish, rodents.

Breeding: varies, mostly 1 off-spring per year; 2-3 offspring in some species; pregnancy lasts from 40 days, sometimes delayed up to 10 months.

Size: smallest (Kitti's hog-nosed bat): head-body 1¼in, weight ¹⁄₂₀ ounce; largest (flying foxes): head-body 16in, weight 3¼lb.

Color: mainly brown, gray and black, tinged red, yellow, orange, silver.

Lifespan: up to 30 years, average about 5 years.

Species mentioned in text:
Common vampire (*Desmodus rotundus*)
False vampire (*Vampyrum spectrum*)
Greater false vampire bat (*Megaderma lyra*)
Greater spear-nosed bat (*Phyllostomus hastatus*)
Hammer-headed bat (*Hypsignathus monstrosus*)
Kitti's hog-nosed bat (*Craseonycteris thonglongyai*)
Large mouse-eared bat (*Myotis myotis*)
Lesser mouse-tailed bat (*Rhinopoma hardwickei*)
Little brown bat (*Myotis lucifugus*)
Mexican long-nosed bat (*Leptonycteris nivalis*)
Natterer's bat (*Myotis nattereri*)
Samoan flying fox (*Pteropus samoensis*)
Schreiber's bent-winged bat (*Miniopterus schreibersi*)

▶A Greater false vampire bat swoops on a mouse. It may take its catch back to its roost to eat it there.

A bat looks rather like a flying mouse. It is the only flying mammal. (The so-called flying squirrel does not fly but just glides, using flaps of skin between its feet.) The bat truly flies, flapping its wings like a bird.

BAD REPUTATION

The Common vampire bat is one of 1,000 species of bat living all around the globe, except in the polar regions. Almost all of the others are harmless and even useful creatures. They help keep down insect pests and vermin and pollinate plants and fruit trees.

But bats in general have had a bad reputation. People have killed them in huge numbers in the past. Today the biggest danger to bat species is the destruction and disturbance of their habitats by human activities.

WINGS AND FINGERS

A bat's wing is a thin membrane or layer of skin, supported by four long

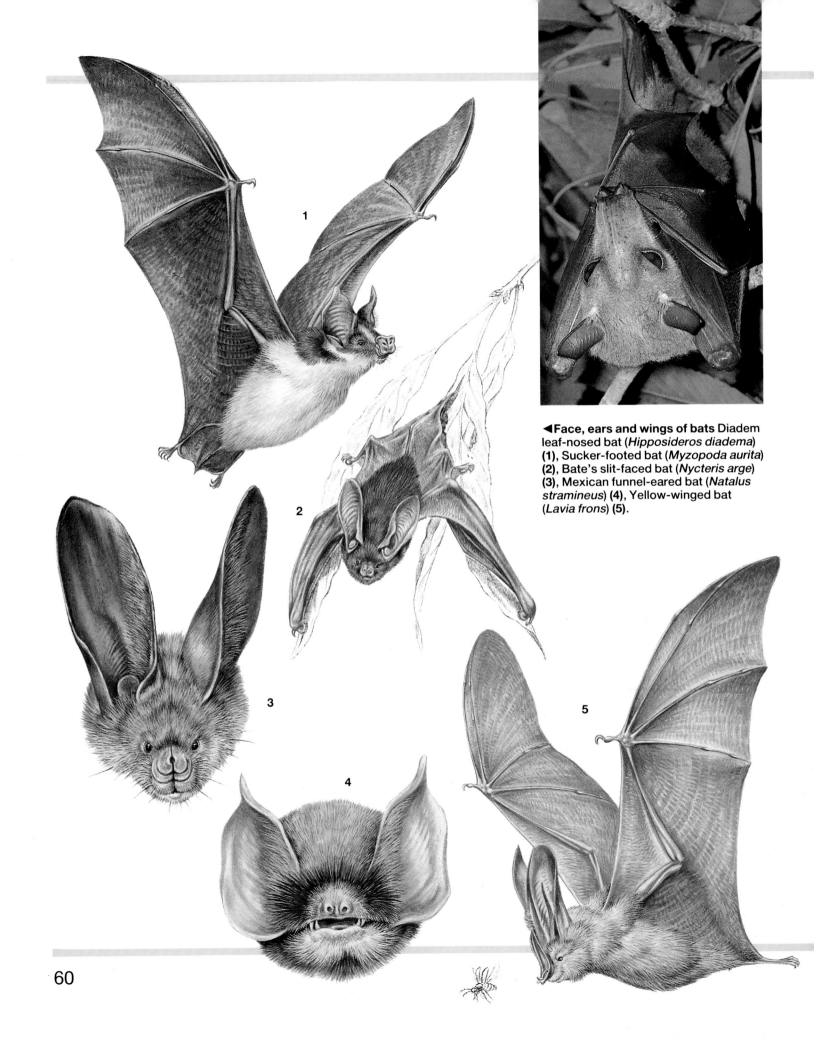

◄**Face, ears and wings of bats** Diadem leaf-nosed bat (*Hipposideros diadema*) **(1)**, Sucker-footed bat (*Myzopoda aurita*) **(2)**, Bate's slit-faced bat (*Nycteris arge*) **(3)**, Mexican funnel-eared bat (*Natalus stramineus*) **(4)**, Yellow-winged bat (*Lavia frons*) **(5)**.

bony fingers. A little clawed thumb protrudes at the front. The thumb is used mainly when the bat is moving around the roost. The wing membrane extends to the bat's legs and, in some species, between the legs to a tail. In other species, such as the Lesser mouse-tailed bat, the tail is free. Some flying foxes have no tail or tail membranes. The bat with the largest wingspan is believed to be the Samoan flying fox. Its wing tips can be as much as 5ft apart.

In most species the legs are weak. The feet have five toes tipped with claws. Bats hang upside-down by their feet when roosting. The Common vampire bat has unusually strong legs. It can run and leap well.

SEEING BY SOUND

Bats sleep during the day in roosting places such as caves, old buildings and trees. They go foraging for food at dusk. Most bats eat insects, which they take on the wing. They have tiny eyes and rather poor eyesight (hence the expression "blind as a bat"). To "see" in the dark, bats rely mainly on sound. They find their way and detect their prey by echo-location or sonar.

When a bat goes in search of insects, it sends out pulses of ultrasound, sound waves too high-pitched for human ears to hear. Sending out about five pulses a second, the bat listens for any echoes. If it receives an echo reflected back from an insect, it increases the pulse rate up to about 200 pulses a second.

►The wing membranes of Davy's naked-backed bat (*Pteronotus davyi*) (1) reach over the back to meet in the middle. The Honduran disk-winged bat (*Thyroptera discifera*) (2). Both species are insect-eaters of the Americas.

From the echoes it receives back, the bat can pin-point exactly where the insect is and snatch it from the air. Usually this is done with the mouth, but the Natterer's bat and some other species first catch insects in their tail membrane.

Bats make these sounds through the open mouth or through the nostrils. The species that use their nostrils have complicated, often grotesque-looking noses. They include the leaf-nosed, hog-nosed, spear-nosed and horseshoe bats. The "leaves" or flaps of skin on the nose help change and focus the sounds coming out.

The ears of bats that use sonar are also unusual. They are large, ridged and folded, which makes them receive the echoes clearly.

FRUIT-EATING BATS

Not all bats navigate and find their food by sonar. Most species of flying foxes use their eyes. These are large bats that feed mainly on fruit; they are often called fruit bats. They eat not only fruit, but also flowers and nectar.

This is how they provide a valuable service in pollinating the plants they visit and spreading the seeds.

Flying foxes look different from the small, odd-faced insect-eaters. They have a long face, large eyes and simple ears, spaced well apart. Their head does look rather like that of a fox. Their eyesight is very good, like that of birds, and they have a strong sense of smell.

Flying foxes are widespread in tropical and subtropical regions of Africa, India, South-east Asia and Australia. Only in the Americas are they absent. There, species of spear-nosed bats are the fruit-eaters.

CARNIVORES AND VAMPIRES

Other species of the spear-nosed bat family are carnivores, eating frogs, rodents and other small animals. Among them is the False vampire, the largest American bat, with a wingspan of up to 40in.

The Americas are also the home of the Common vampire. This bat's range extends from Mexico southwards to Argentina. The Common vampire feeds mostly on the blood of domesticated livestock, especially large herds of cattle. Attacks on humans are very rare. It takes only a little of an animal's blood and so does not harm it in that way. But the Common vampire can infect animals with diseases, including the deadly rabies.

BAT COLONIES

Bats feed mostly by night and roost by day. Flying foxes roost in the open, hanging from the branches of trees. Some bats nest in holes in trees, others in buildings, canal tunnels and other human constructions. The largest numbers of all roost in caves. Some colonies of cave-dwelling bats contain a million individuals or more.

Caves are ideal roosting places. They are safe and dark and have a steady temperature. With their skill at echo-location, bats have no problem finding their way around inside.

Among most species males and females roost together for much of the year. But some species roost in single-sex groups. There is not much pattern in the way bats mate, although there are a few exceptions. The Greater spear-nosed bat forms harems of females, with just one male mating with them.

Mating takes place at roost. Sometimes pregnancy is delayed until the climate is more suitable, maybe as long as 7 months after mating. Few other mammals do this. The young are sometimes born with their mother hanging upside-down. Otherwise the mother turns her head upwards and then catches the young in her tail membrane. The young begin suckling their mother almost immediately. In most bat colonies all the mothers and young roost together in a kind of nursery.

BEATING THE CLIMATE

In parts of the world where the climate is always warm, bats usually

stay active all the year. In the tropics bats may sleep through the hottest parts of the year. This period of sleep is called estivation.

In cooler climates, bats have two main problems in winter. Their supply of food (notably insects) decreases; and the temperature falls. To cope with this, some bats migrate, making their way to warmer climates.

▲Looking like a whirlwind, a huge colony of bats leave a cave in Java to feed at dusk.

▶Red and naked, hundreds of newly born Schreiber's bent-winged bats on the roof of a nursery cave in Australia. They will remain in the cave for about 3 months.

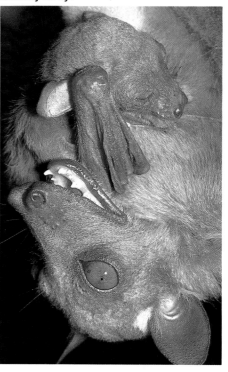

◄A colony of fruit bats roosting during the daytime, their wings wrapped around them. In hot weather, they open out and flap their wings to keep cool.

▼Flying fox mother with newly born young. Three months may pass before it is ready to fly.

Others hibernate, sleeping through the winter cold.

Before they hibernate, these bats feed well to build up their body weight, including a thick layer of fat under the skin. Bats usually cluster together when they hibernate, which helps keep them warm. This can result in huge clusters of bats packed tightly together, with as many as 3,000 in a square yard. The Little brown bat of North America and the Large mouse-eared bat of Europe may hibernate together like this.

Hibernating bats do not spend the whole winter asleep. Every 10 days or so they wake up and sometimes fly a short distance to another site. This probably helps rid their bodies of wastes which build up and could otherwise poison them.

63

ECHIDNAS

A female echidna is resting quietly in her burrow. Inside her pouch is a small, soft, leathery egg, which she laid 10 days ago. The egg starts to move and soon splits open. A tiny form no bigger than a peanut crawls out. It makes its way to the milk glands in the mother's pouch and starts to suckle.

The echidna is one of the most unusual of all mammals because it lays eggs. The only other mammal to do this is the platypus. Both these mammals are called monotremes, a word meaning "one hole". They have a single opening, not two, at the rear of the body.

SPINY COAT

There are two species of echidna. The smaller species, with the shorter snout, is the Short-beaked or Common echidna. It looks much like a hedgehog. It has long sharp spines sticking out of thick dark fur. The spiny coat hides a short tail and also covers ear-slits just behind the bulging eyes.

The Short-beaked echidna is found throughout Australia and Tasmania and also on the island of New Guinea to the north. It lives in almost all kinds of habitat, from dry desert to rain forest and snowy mountains.

The Long-beaked echidna is also found in New Guinea, but only in highland regions. It has longer fur than its Short-beaked relative. It also has fewer and shorter spines, which are usually visible only at the sides and on the head.

In both species the males are heavier than the females. Males have a horny spur on the ankle of the hind limbs, which they use when fighting.

DIET OF ANTS AND WORMS

Echidnas are often called spiny ant-eaters. But only the Short-beaked species eat ants and termites. The Long-beaked echidna feeds mainly on earthworms. Like the anteaters, however, echidnas have no teeth. The Short-beaked echidna takes ants and termites with its long tongue, made

ECHIDNAS Tachyglossidae
(*2 species*)

Habitat: semi-desert to highlands.

Diet: ants, termites, earthworms.

Breeding: young hatched from 1 egg after about 10 days incubation.

Size: Short-beaked echidna: head-body 12in, weight 5½lb; Long-beaked echidna: head-body 36in, weight 22lb.

Color: black to brown coat, with paler-colored spines.

Lifespan: up to 50 years in captivity.

Species mentioned in text:
Long-beaked echidna (*Zaglossus bruijni*)
Short-beaked or Common echidna (*Tachyglossus aculeatus*)

▲ The long naked snout of the Long-beaked echidna curves downwards. The mouth at the tip is tiny. It can be opened just wide enough to allow the tongue to pass through.

► When swimming, the Short-beaked echidna uses its snout as a snorkel, pushing the tip into the air to breathe.

▲The nostrils of a Short-beaked echidna are much larger than its mouth. The eyes are bulging. The powerful forelimbs are tipped with tough claws for digging.

sticky with saliva. The insects are crushed between spines at the back of the tongue and the roof of the mouth. The Long-beaked echidna catches earthworms in spines that run in a groove in its tongue. It then draws the worms into its tiny mouth, head or tail first.

►With only its spines showing, a Short-beaked echidna digs into an ants' nest. In soft soil echidnas may bury themselves to shelter from the Sun's heat.

PLATYPUS

The dog sniffing around near the river bank suddenly gives an excited bark. It makes a short dash and pounces on a small furry creature, which has the bill of a duck and the flat tail of a beaver. This odd-looking creature is a platypus – and unfortunately for the dog, it's a male. The platypus wriggles this way and that and then manages to jab the dog with the spurs on its hind legs. They deliver a powerful venom. Startled and in pain, the dog drops the platypus, which escapes into the river.

The platypus is the only mammal besides certain shrews that is venomous. But only the male is able to produce and deliver the poison, which can kill a dog and cause agonizing pain to human beings.

The platypus is also most unusual among mammals because it reproduces by laying and hatching eggs. When the young hatch, they feed on their mother's milk, like all mammals do. The only other mammals to lay eggs are the echidnas, which together with the platypus make up the animal order of monotremes.

The platypus is found in eastern Australia. It lives in burrows in the banks of rivers and lakes, spending much time in water. Once hunted nearly to extinction for its fur, the platypus, now protected, is thriving.

SLEEK SWIMMER
The platypus moves awkwardly on the land on its short legs. But in the water it is swift and graceful. Its body, covered with short thick fur, becomes beautifully streamlined, and it propels itself with its broad webbed forefeet. It steers with its partly webbed hind feet and flat tail.

When walking on land, the platypus

folds back the webbing on its forefeet. This exposes thick nails, which the platypus uses for digging its burrows.

FEEDING AND BREEDING
When the platypus dives into water, it closes its eyes and ears. Under water the platypus's soft, rubbery and skin-covered bill becomes its eyes and ears. The bill is very sensitive to touch, which helps the animal find its food as

▼ The female platypus digs a breeding burrow where she will lay her eggs and raise her young. The young suckle her and remain in the burrow for up to 4 months.

▲ The platypus is often called the duck-bill. But unlike a duck's bill the bill of the platypus is soft and flexible.

PLATYPUS *Ornithorhynchus anatinus*

〰 Habitat: rivers and lakes.

■ Diet: insect larvae, worms, crustaceans.

◯ Breeding: young hatched from usually 2 eggs after 10 days' incubation.

Size: head-body 18-24in, bill 2in, tail 4-6in, weight 2-6lb male; female somewhat smaller.

Color: dark brown coat on back, silvery to light brown underneath.

Lifespan: up to 17 years in captivity.

▲ The platypus feeds mainly on the river bottom. It uses its sensitive bill to sift through the mud and gravel for insect larvae and small shellfish.

it searches for food on the river bed.

The platypus scoops up insect larvae and crustaceans in its bill, then stores them in its two cheek pouches, located just behind the bill. The platypus has no teeth, but inside the cheek pouches are horny ridges which help grind the food into smaller pieces.

It is thought that platypuses mate in the water after the female slowly approaches the male and he then chases her and grasps her tail. Some days later the female starts to dig a long breeding burrow. She makes at the end a cozy nesting chamber, lined with grass and leaves. There she lays up to three soft, leathery eggs and keeps them warm.

In about 10 days the young hatch and crawl to the mammary glands on the mother's belly and start to suck the milk-soaked fur there.

OPOSSUMS

OPOSSUMS Didelphidae
(*75 species*)

Habitat: mainly wooded areas in temperate and tropical regions.

Diet: grass, fruit, insects, snakes, birds, other small animals, carrion.

Breeding: many young after pregnancy of about 2 weeks; up to about 10 young successfully raised.

Size: smallest (Formosan mouse opossum): head-body 2in, tail 1½in, weight ⅓ ounce; largest (Virginia opossum): head-body 22in, tail 22in, weight 12lb.

Color: gray, brown and golden coat, sometimes striped.

Lifespan: up to about 3 years, longer in captivity.

Species mentioned in text:
Brown four-eyed opossum
   (*Metachirus nudicaudatus*)
Formosan mouse opossum
   (*Marmosa formosa*)
Little water opossum or lutrine
   (*Lutreolina crassicaudata*)
Southern opossum (*Didelphis marsupialis*)
Virginia or Common opossum
   (*D. virginiana*)
Water opossum or yapok
   (*Chironectes minimus*)
Woolly opossum (e.g. *Caluromys lanatus*)

A fox is on the prowl, nose to the ground, following the scent of an animal which could be its next meal. Then it sees its prey, a long-nosed, long-tailed furry creature as big as a cat. It is a Virginia opossum. The fox lunges at the animal and snatches it up in its teeth, shaking it from side to side. The opossum hangs limply in its jaws, eyes closed and tongue hanging out. The fox tosses the opossum to the ground where it lies motionless, as if dead. The fox is confused as the animal did not struggle, and moves off in search of live prey. The "dead" opossum then runs away.

"Playing possum," or appearing to be dead, is one way the Virginia opossum escapes the attention of predators. The Virginia opossum is the only one to do this and the only opossum to be found in North America. The many other species of opossum are found from Mexico throughout Central and South America.

Opossums belong to the order of animals known as marsupials. They are mammals which usually raise their young in pouches. Most marsupials live in Australia, including the kangaroo and the koala. Opossums are the only marsupials found outside Australasia.

EXPERT CLIMBERS
Many species of opossum spend much of their life in the trees, and their bodies are well adapted for climbing. Their feet have sharp claws and the big toe on the hind foot is "opposable". This means it can act opposite the other four toes to make grasping easier.

Most opossums also use their long, prehensile tail as an extra limb for grasping. An opossum can curl its tail around a branch to help steady itself or hang upside-down.

The woolly and four-eyed opossums inhabit the humid tropical rain forests of South America. The Woolly opossum has the large, bulging eyes of a typical tree dweller. It usually stays in the upper canopy of the forest, feeding on nectar and fruit.

Four-eyed opossums are usually found in the lower branches of the

forest or on the ground. There they feed on insects, worms, other animals and fruit. They get their name from the white spots on their forehead. The Brown four-eyed opossum is sometimes called the rat-tailed opossum,

▶**Species of opossum and relatives**
Red-sided short-tailed opossum (*Monodelphis brevicaudata*) eating a centipede (1). Brown four-eyed opossum grooming (2). White-eared opossum (*Didelphis albiventris*) hanging by its tail (3). Ashby mouse opossum (*Marmosa cinerea*) climbing (4). Yapok catching fish (5). Lutrine being aggressive (6). Woolly opossum (7). Patagonian opossum (*Lestodelphys halli*) hunting a spider (8). Gray four-eyed opossum (*Philander opossum*) eating fruit (9). Black-shouldered opossum (*Caluromysiops irrupta*) eating nut (10). Bushy-tailed opossum (*Glironia venusta*) climbing (11). Opossum relatives: Monito del monte (*Dromiciops australis*) in nest (12) and Shrew opossum (*Lestoros inca*) feeding (13).

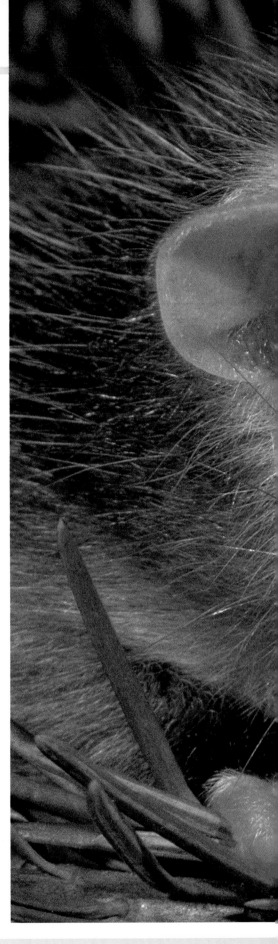

and it does look very much like a rat.

The most numerous of opossum species are the mouse opossums, of which there are nearly 50. They are so called because of their size. Some live in forests, others in more open country and grassland.

The even smaller short-tailed opossums are also found in more open countryside. These mainly ground dwellers look much like shrews.

"WATER WEASELS"

Two species of opossum are good swimmers. One is the Little water opossum or lutrine. Like the weasel, which it resembles, it is a fierce carnivore, feeding on small mammals, birds and frogs. It is found along rivers in open country and even in suburban areas.

The Water opossum or yapok is even better adapted to life in the water. It has webbed feet, and the female can close its pouch under water to protect the young inside. It too is carnivorous.

IN THE POUCH

The female opossum gives birth about 2 weeks after mating. The young are

▲Virginia opossum mother and young. They are about 3 months old and still hitch a ride on their mother.

▶An alert Southern opossum sniffs the air. Its ears move to catch the faintest sounds that may tell it that a juicy meal is nearby.

poorly developed, and no bigger than a honeybee. Blind and naked, they claw their way from the birth canal into the mother's furry pouch. There, if they are lucky, they attach themselves to a nipple and start to feed. They remain in the pouch for about 2 months and then start crawling on the mother's back.

The female opossum often produces more offspring than she has nipples. This means that some of the young die. The Virginia opossum has been known to produce over 50 offspring but has only about 13 nipples.

Some opossums, including the mouse and short-tailed opossums, have no pouch for breeding. The young attach themselves to the mother's nipples and dangle from them when she moves. If they fall off they let out a high-pitched cry, calling the mother back.

BANDICOOTS

Two male bandicoots patrolling neighboring territories come face to face. On other nights one has chased the other away. Tonight both stand their ground, and a fight breaks out. Up on their hind legs, they lunge and lock jaws, wrestling each other to the ground. One struggles free, leaps high into the air and rakes the other with its hind claws. Squealing in fright, the other runs away. The victor gives chase for a while, then resumes the nightly patrol.

Bandicoots are stocky, coarse-haired mammals with a pointed snout. They are the size of a rabbit. They have the small, even teeth of a typical insect-eater. Their feet have strong claws,

▶ The white markings on the rump give the Eastern barred bandicoot its name.

▼ Some typical postures of a short-nosed bandicoot. Squatting to sniff the air when hunting (1). Male hopping aggressively (2). Digging for insects with strong foreclaws (3). Female giving birth (4). Mother carrying 7-week-old young in pouch (5).

which they use for digging into the ground for insects and worms.

Bandicoots live in Australia and New Guinea and on neighboring islands. Like many Australian mammals, they are marsupials, raising their young in pouches. The pouch opens at the rear, not at the front as with other marsupials.

THE LONG AND THE SHORT
The hind limbs of the bandicoot are longer than the forelimbs, especially in the Greater bilby. This increasingly rare animal lives in the Australian deserts. Unlike other bandicoots, it has very long ears and a furry tail.

The common Brindled bandicoot has short ears and a nearly naked tail. It lives in the wetter regions near the coasts of eastern and northern Australia and further inland along rivers. The Brindled bandicoot is one of the short-nosed bandicoots, which gener-

BANDICOOTS Perameli-dae, Thylacomyidae (*21 species*)

● ◪ 🐾

◧ Habitat: desert to rain forest.

◩ Diet: insects, worms, other small animals, fruit and seeds.

◎ Breeding: up to 7 young after pregnancy of 2 weeks or less.

Size: smallest (Mouse bandicoot): head-body 6in, tail 4in, weight 1 ounce; largest (Giant bandicoot): head-body 22in, tail 13in, weight 10lb.

Color: gray or brown coat.

Lifespan: up to 5 years.

Species mentioned in text:
Brindled bandicoot (*Isoodon macrourus*)
Eastern barred bandicoot (*Perameles gunnii*)
Giant bandicoot (*Peroryctes broadbenti*)
Greater bilby or Rabbit-eared bandicoot (*Macrotis lagotis*)
Long-nosed bandicoot (*Perameles nasuta*)
Mouse bandicoot (*Microperoryctes murina*)

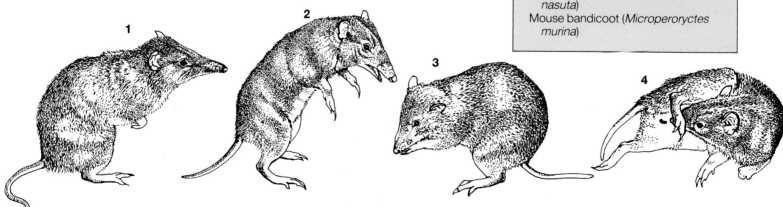

▲A female Greater bilby and young. The animal's alternative name, Rabbit-eared bandicoot, describes it well.

5

ally live in long grass and shrub where there is good ground cover. The long-nosed bandicoots prefer more open ground such as short grassland. They have longer ears as well as longer muzzles.

The long-nosed bandicoots of the grasslands, such as the Eastern barred bandicoot, have striped or barred markings on the coat to provide camouflage. Such markings are absent in the forest species.

FAST REPRODUCTION

Male bandicoots patrol their home territory each night to find females ready to mate and to keep out rival males. This is when they may fight.

The female gives birth usually in less than 2 weeks. Unlike those of other marsupials, the young are attached to the mother at birth by a placenta, as most mammals are. The young crawl into the mother's pouch and attach themselves to a nipple and begin to feed. They remain in the pouch for about 7 weeks and begin to eat normal food about 10 days later.

By this time the female may be pregnant again or may already have another litter in the pouch. But despite this high rate of reproduction, the population of Australian bandicoots has fallen rapidly in recent years. The desert species have been especially reduced. Overgrazing by cattle, sheep and rabbits is probably one of the main causes.

POSSUMS AND GLIDERS

POSSUMS AND GLIDERS Pseudocheiridae, Burramyidae, Petauridae (*30 species*)

Habitat: forest, scrub, heath.

Diet: leaves, gum, nectar insects.

Breeding: 1 or 2 (ringtail possums) or 4-6 (pygmy possums) young after pregnancy of 2-4 weeks.

Size: smallest (Little pygmy possum): head-body 2½in, tail 3in, weight ¼ ounce; largest (Rock ringtail possum): head-body 15in, tail 11in, weight 4½lb.

Color: gray or brown coat, paler underneath, often darker eye patches.

Lifespan: up to 15 years.

Species mentioned in text:
Common ringtail possum
(*Pseudocheirus peregrinus*)
Feathertail or Pygmy glider (*Acrobates pygmaeus*)
Greater glider (*Petauroides volans*)
Striped possum (*Dactylopsila trivirgata*)
Sugar glider (*Petaurus breviceps*)
Yellow-bellied or Fluffy glider
(*P. australis*)

High in the eucalypt forest a long-tailed animal leaps from a branch into the air. It is not aiming for another branch nearby but for a tree trunk over 100ft away. As it becomes airborne it spreads its arms and legs wide, stretching out a flap of skin between wrists and ankles. Gracefully and effortlessly the animal glides and maneuvres through the trees to make a perfect landing, right on target.

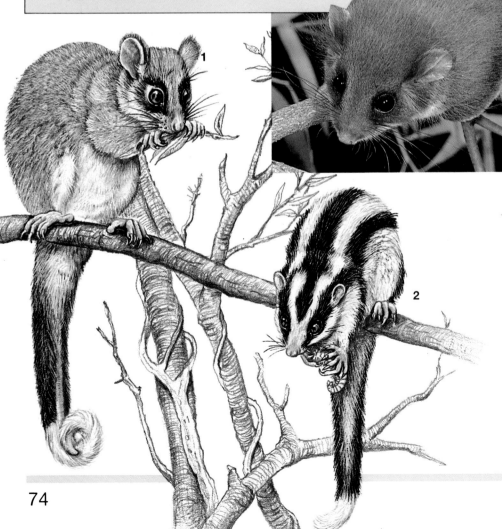

▲ The Feathertail glider of eastern Australia is one of the several species of pygmy possums.

► The Sugar glider (1) and Yellow-bellied glider (2) are both expert gliders. They feed on sweet sap and gum. The Yellow-bellied glider makes V-shaped notches in the tree trunks when it feeds.

◄ A Common ringtail possum curls its tail as it eats its favorite leaves (1). The Striped possum eats an insect (2).

DICK TWINNEY 84

Animals that glide are common in the forests of Australia, New Guinea and the nearby islands. They include the Yellow-bellied or Fluffy glider of eastern Australia which can "fly" up to 330ft. All the Australian and New Guinea gliders belong to three related families: the ringtail possums, the gliders and the pygmy possums.

Australian possums were so named because they looked like the American opossums. Like opossums they are marsupials – after birth, their tiny naked young crawl into the mother's pouch, attach themselves to a nipple and suckle for several weeks.

GLIDING AND GRIPPING

There are nine species that glide, using a membrane of skin between their limbs. When flying, the long tail of a glider stretches out behind. This helps steady the flight and aids steering. Most gliders have a thick furry tail, but the Feathertail glider's tail has tufts of hair on the end which look like the flight feathers of an arrow.

The gliders have long and sharp claws, especially on their forefeet. This helps them grip the bark of the trees when they land. Both gliders and non-gliders have large hands and feet with digits that can be spread apart to get a firm grip.

The non-gliders have a prehensile tail with which they can also grip. The tail is usually naked underneath to get a firmer hold. The ringtail possums have the habit of curling up the end of the tail when they are not using it. This is how they got their name.

VARIED DIETS

The ringtail possums and the closely related Greater glider eat mainly leaves. The Yellow-bellied and Sugar gliders mostly feed on the sap and gum of trees such as wattle and eucalyptus.

The Striped possum eats insects. It has large teeth, a long tongue and a very long fourth finger to help get at insects and larvae beneath tree bark. It is particularly partial to eating ants, bees, termites and other wood-boring insects. As it picks away at the bark, it produces a shower of wood-chips. The mouse-like pygmy possums prefer nectar and pollen. This rich diet makes them breed bigger litters (up to 7) than the other possums and makes them grow more quickly.

1

2

AARDVARK

The Sun has set over the sparsely wooded grassland. Dotted here and there, nearly as tall as the trees, are the great mounds of termite nests. An animal the size of a pig makes its way in a zigzag path towards one of the mounds. It is an aardvark on its way to feed. A hole at the base of the mound shows it has been there before. It sets to work again, snout close to the ground, digging swiftly. It makes good progress, even though the mound is rock hard. From time to time the aardvark pauses, sits down and thrusts its snout into the mound to feed on the insects scurrying about inside.

◄ An aardvark sits down by a termite mound ready to feed at night. Termites in the wet season and ants in the dry season are its favorite foods.

▼ In parts of Zaire, people kill aardvarks both for food and for their hair and teeth. The hair is crushed into a powder and used as a poison. The teeth are used as good-luck charms.

Aardvarks are found in most parts of Africa south of the Sahara desert. But they avoid the driest desert regions and the depths of the tropical rain forest. The name aardvark means "earth pig" in the Afrikaans language, and the animal does look rather like a pig, although it is no relation. It has a long round snout, big ears and a bristly coat.

The aardvark eats ants and termites, but it is not related to anteaters. In fact it is not closely related to any other mammal. Its nearest relatives appear to be hyraxes and elephants. It has certain body features in common with these animals – for example, their claws are a cross between nails and hoofs. For this reason they are sometimes classed together as "primitive ungulates". Ungulates are the hoofed mammals.

THICK-SKINNED AND SHY
The aardvark eats ants and termites using its long, worm-like tongue, which is sticky with saliva. Its skin is thick and tough, which protects it from insect bites.

The aardvark has keen hearing and a well-developed sense of smell, senses it uses when foraging for food. A mat of hair around the nostrils helps keep out the dust. The animal can also close

its nostrils when digging to prevent dirt and insects getting in.

Aardvarks are shy and secretive creatures, whose way of life is not well known. They spend the day asleep in a tunnel-like burrow. They dig several burrows within the home territory and may rest in a number of them during the night.

The young are born in a deeper and longer burrow system. They emerge above ground after about 2 weeks and go out foraging with their mother. They begin to dig their own burrows at 6 months old, but usually stay with the mother until she mates again.

AARDVARK
Orycteropus afer

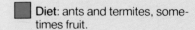

Habitat: open woodland, scrub and grassland.

Diet: ants and termites, sometimes fruit.

Breeding: 1 young born after pregnancy of about 7 months.

Size: head-body 40-52in; tail 18-25in; weight 88-145lb.

Color: yellowish-gray coat, head and tail off-white, often stained brownish by soil.

Lifespan: up to 10 years in captivity.

ELEPHANT-SHREWS

ELEPHANT-SHREWS

Macroscelididae (*15 species*)

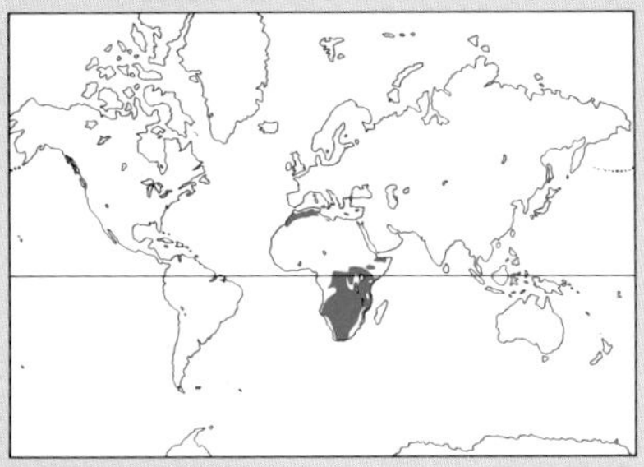

Habitat: from forest to desert.

Diet: insects, spiders, worms.

Breeding: usually 1 or 2 offspring after pregnancy of 40-65 days.

Size: smallest (Short-eared elephant-shrew): head-body 4in, tail 4½in, weight 1½ ounces; largest (Golden-rumped elephant-shrew): head-body 12in, tail 10in, weight 1¼lb.

Color: gray, light or dark brown coat, with streaks or patches.

Lifespan: up to 4 years.

Species mentioned in text:
Black and rufous elephant-shrew (*Rhynchocyon petersi*)
Golden-rumped elephant-shrew (*R. chrysopygus*)
Rufous elephant-shrew (*Elephantulus rufescens*)
Short-eared elephant-shrew (*Macroscelides proboscideus*)

From its perch, the eagle spots the eye-catching rear of the Golden-rumped elephant-shrew some distance away. It turns its head and shuffles its legs, preparing to attack. But the elephant-shrew has seen the eagle move and knows it has enough time to escape. It slaps the ground noisily with its tail as if to say, "I've seen you. You can't catch me!"

The elephant-shrews are insect-eating mammals that live in most parts of Africa except the west. They have a long snout like an elephant's trunk and a similar diet to shrews – hence their name. But they are quite unrelated to shrews (or elephants!). They are much bigger, with big ears, large bulging eyes and very long legs, almost like a miniature antelope.

There are two main groups of elephant-shrews: giant ones, such as the Golden-rumped elephant-shrew, and small ones, such as the Rufous and Short-eared elephant-shrews. The small ones are approximately mouse-sized, the large ones bigger than rats.

▼**Four species of elephant-shrews** North African elephant-shrew (*Elephantulus rozeti*) washing its face **(1)**. Short-eared elephant-shrew clearing a trail **(2)**. Four-toed elephant-shrew (*Petrodamus tetradactylus*) probing for insects with its tongue **(3)**. Checkered elephant-shrew (*Rhynchocyon cirnei*) scent-marking with its tail glands **(4)**.

TAKING FLIGHT

The elephant-shrew's long legs make it a very fast mover. If the Golden-rumped elephant-shrew wants to escape from predators, it does so in leaps and bounds at speeds of 15mph.

Some of the smaller species, including the Rufous elephant-shrew, make trails through their territory. They keep these trails free from debris so that they can when necessary run along them at top speed. It could make the difference between life and death if they were chased.

KEEPING IN TOUCH

Elephant-shrews generally live in mating pairs. Both sexes are similar in

▼A Rufous elephant-shrew forages for insects (1). It lives for only about 2½ years in the wild. The giant Black and rufous elephant-shrew devours a centipede, one of its favorite foods (2).

▲The conspicuous patch on the Golden-rumped elephant-shrew could be a target for attackers. But the animal has a good chance of surviving these attacks because the skin is very thick there.

appearance, and share the work of maintaining their trails. They also both defend their territory against intruders. The males fight off rival males, the females fight rival females.

The pair do not spend all their time together, but they keep in contact around their home territory. They may do so by scent-marking, by slapping on the ground with their tail or by drumming with their hind feet.

Young elephant-shrews are born very well developed, with a furry coat and sometimes with their eyes already open. They are usually able to leave the nest with their mother in a day or so. In parts of Kenya especially, many fall victim to hunters, who snare and eat them. The forest-dwelling giant elephant-shrews are threatened by destruction of their homes.

TREE SHREWS

Up in the branches of the rain forest a tree shrew seems to be having fun. It is sliding backwards down a branch with its body pressed against the bark. But this is no game the tree shrew is playing. It is marking the branch with scent from a gland in its abdomen. This activity, called sledging, helps mark its territory.

Scent-marking plays an important role in the life of tree shrews. These small squirrel-like mammals live in the tropical rain forests of India and South-east Asia. They are not well named, as they do not look much like shrews, or always live in trees.

Tree shrews are unusual because they have features in common with primitive primates (such as tarsiers) as well as with insectivores like shrews. But they are now usually classed in a completely separate order of animals (Scandentia). Primates, insectivores and tree shrews probably evolved from the same ancestors many millions of years ago.

THE LONG AND THE SHORT
All tree shrews have a relatively long soft furry body. In most species the long tail is quite bushy. But the tail of the Northern smooth-tailed tree shrew of Vietnam, Cambodia and Thailand is only sparsely covered. And that of the Pen-tailed tree shrew of Malaysia, Sumatra and Borneo is naked except for a tuft at the tip.

The Pen-tailed and smooth-tailed tree shrews spend nearly all their time in the trees. They have a short snout, but the tail is very long. This helps them keep their balance when aloft. The Pen-tailed is the only tree shrew active at night.

Tree shrews that spend more time on the ground, such as the Terrestrial tree shrew, have a shorter tail. They are also bigger and have longer snouts, which they use for rooting among the leaf litter.

BAD MOTHERS
Female tree shrews take less care of their young than almost any other mammal. After birth they leave the young in a brood nest, while they sleep in another. They visit the young briefly only once every 2 days to feed them. During this time the young gorge themselves full and become bloated.

The number of offspring varies from species to species and can be gauged from the numbers of pairs of teats the female has. Belanger's tree shrew has three pairs and can raise three offspring. These grow rapidly and can reach maturity in only 4 months, so that if food is plentiful the animals quickly colonize new areas.

►Activities of some species of tree shrew. Pen-tailed tree shrew eating a beetle (1). Common tree shrew "chinning" to leave a scent trail (2). Northern smooth-tailed tree shrew snatching an insect from the air (3). Terrestrial tree shrew turning over a stone to look for prey (4).

▲Tree shrews eating an earthworm. They usually hold their food with their forepaws when they feed.

►This Common tree shrew from South-east Asia is devouring an insect. It has a medium-sized snout and spends only part of its time in the trees.

TREE SHREWS Tupaiidae
(*18 species*)

Size: head-tail 8-12in, weight 1½-2 ounces.

Color: grayish, brown or reddish coat, whitish or orange underneath, often pale shoulder stripes and facial streaks.

Lifespan: not known.

Species mentioned in text:
Belanger's tree shrew (*Tupaia belangeri*)
Common tree shrew (*T. glis*)
Northern smooth-tailed tree shrew (*Dendrogale murina*)
Pen-tailed tree shrew (*Ptilocercus lowii*)
Terrestrial tree shrew (*Lyonogale tana*)

Habitat: mainly rain forest.

Diet: mainly insects, worms, also some fruit and seeds.

Breeding: up to 3 young after pregnancy of 45-50 days.

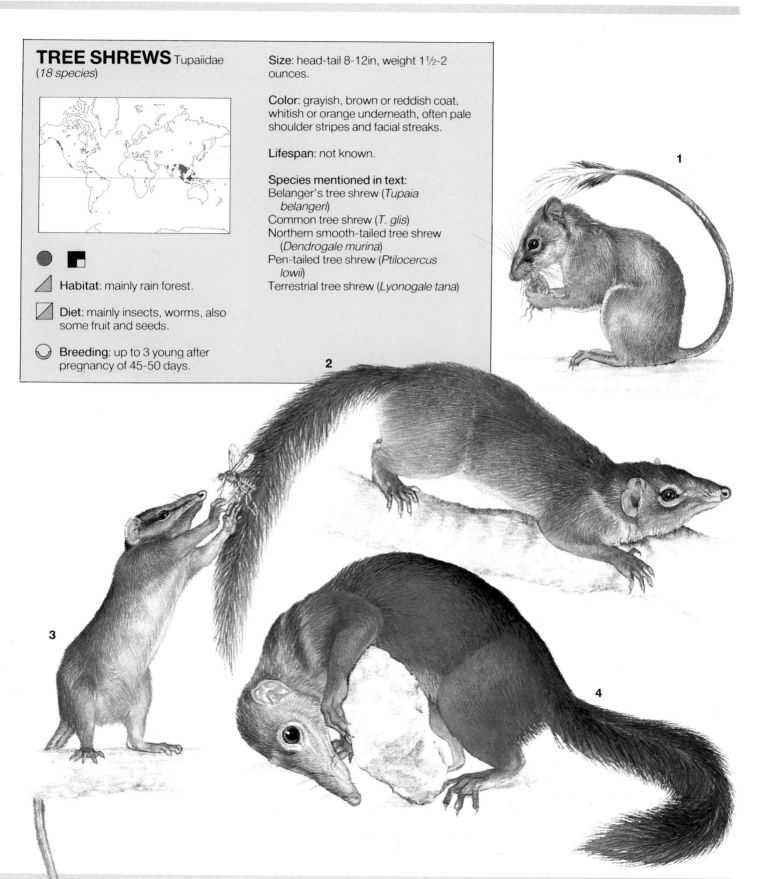

GRAY WHALE

With a flick of its huge tail, the Gray whale power-dives its way down to the sea bed. There it swims along the bottom, its head ploughing through the thick layer of sediment. It scoops up a mouthful of crustaceans, worms and muddy water and returns to the surface. On the surface it expels the muddy water through its plates of baleen and swallows the food filtered out. Then it dives again.

▼Gray whale mother and calf. The calf is especially sleek and smooth-skinned.

GRAY WHALE *Eschrichtius robustus*

■ Diet: bottom-dwelling marine animals, such as crustaceans and worms.

◎ Breeding: 1 offspring after pregnancy of about 13 months.

Size: head-tail 39-47ft, weight 17½ tons male; head-tail 42-50ft, weight (pregnant) up to 37½ tons female.

Color: mottled gray.

Lifespan: up to 77 years.

○ ■ 🐋

Habitat: mainly coastal waters.

◀Large clusters of barnacles cover much of the Gray whale's skin as it gets older. Pale spidery whale lice live in the barnacle clusters.

▲The Gray whale blows as it comes up to the surface for air (1). Soon it dives again (2). Sometimes it "spy-hops" to get its bearings (3).

The Gray whale is one of the baleen whales. These whales feed on small sea creatures, which they catch by straining mouthfuls of sea water through horny plates called baleen (whalebone). The Gray whale's baleen is much shorter and stiffer than that of the other baleen whales, such as the Blue and Right whales. Like all the whales, it is a mammal whose young are, from birth, raised on the mother's milk.

There are two main populations of Gray whales, which migrate between Arctic and Southern Pacific ocean waters. The western or Korean population migrates between Siberia and South Korea. This population is probably almost extinct.

The eastern or Californian population migrates between the Bering Sea and the southern coast of California. It has recovered well from a devastating century of whaling, which ended in the 1940s. Today the population numbers are estimated to be more than 15,000.

CALIFORNIAN MIGRATION
The Californian Gray whales make an annual migration of up to 12,000 miles. They keep quite close to the North American coast in waters less than 330ft deep.

In the summer the whales feed in the nutrient-rich waters of the Bering Sea. They put on weight fast, building up a thick blubber layer beneath the skin. They need this blubber as an energy store, because when they migrate they eat very little.

They start moving south in the fall, reaching southern California by December. There they mate. By March they are heading north again for their summer feeding grounds.

The pattern is repeated each year. In December, in the seas around California, the females that became pregnant the previous year give birth.

▲With most of its gigantic body submerged, a Gray whale blows, expelling stale air from its lungs. It may do this up to five times after surfacing.

NURSING MOTHERS
The whale young, or calf, measures nearly 16½ft at birth. It has difficulty in breathing and swimming at first, and often the mother has to support it on the surface with her back or tail fins. The mother's teats are hidden, but they spurt milk into the calf's mouth when it nuzzles the right spot.

The calves stay close to their mothers on the long migration north in the spring. By the time they reach the northern feeding grounds they have become skilful swimmers and have built up a thick insulating layer of blubber.

RORQUALS

The 66-ton bulk of a Humpback whale surges upwards through the water, driven by thrusts of its powerful tail. It breaks the surface and arches into the air, then falls back with a loud smack that makes the sea boil. No other whale performs this action, called breaching, so acrobatically.

RORQUALS
Balaenopteridae (*6 species*)

■ Diet: plankton, krill, fish.

◎ Breeding: 1 offspring after pregnancy of 10-12 months.

Size: smallest (Minke whale): head-body 36ft, weight 11 tons; largest (Blue whale): head-body 89ft, weight 165 tons.

Color: black or blue-gray, often white underneath.

Lifespan: 45 years (Minke whale) to 95 years (Humpback whale).

○ ■ 🏴

▨ Habitat: all main oceans.

Species mentioned in text:
Blue whale (*Balaenoptera musculus*)
Fin whale (*B. physalus*)
Humpback whale (*Megaptera novaeangliae*)
Minke whale (*Balaenoptera acutorostrata*)
Sei whale (*B. borealis*)

1

2

The family of whales known as the rorquals includes the largest animal that has ever lived on this planet, the Blue whale. When fully grown, the Blue whale weighs more than 30 African bull elephants and is as long as eight automobiles placed bumper to bumper. The Fin whale grows almost as long, although it is slimmer and only about half the weight. Then in order of bulk come the Sei, Humpback, Bryde's and Minke whales.

The great stocks of these whales that once lived in the oceans have been devastated by whaling. Thousands were killed each year. Now they are protected, but they are still vulnerable to other hazards, such as pollution.

▼**Species of rorqual** Five of the six rorquals. Humpback whale "breaching" **(1)**. Smallest of the rorquals, the Minke whale **(2)**. Bryde's whale (*Balaenoptera edeni*) **(3)**. Blue whale **(4)** – the slightly smaller Pygmy blue whale (*B. musculus brevicauda*) is found in southern waters. Fin whale **(5)**, which may have most baleen plates and throat-grooves.

▲The head of a Blue whale breaks the surface. It is sucking in air through the blowholes, ready for its next dive into the deeps.

FILTER-FEEDERS
Like all whales, the rorquals are mammals. They are filter-feeders,

straining their food from the water using comb-like plates of baleen (whalebone) in their upper jaws. First they take a mouthful of water containing food and then force the water out through the baleen. The food gets trapped by bristles on the plates and is then swallowed.

All the rorquals have folds or grooves in the throat extending back along the belly. They allow the whales to increase enormously the volume of their mouths when gulping water to feed.

The favorite food of the rorquals is krill, a shrimp-like creature only a few inches long. An adult Blue whale can eat as much as 2½ tons of krill every day.

Smaller plankton called copepods and fish such as herring, mackerel and cod are also part of the rorquals' diet. The water-thumping behavior of breaching may be one way in which whales help scare and concentrate the fish before they feed.

GRACE AND PACE
Despite their huge bulk, the rorquals are graceful swimmers. Their bodies are beautifully streamlined, and they propel themselves by up and down movements of their large tail fins. The Sei whale is probably the fastest swimmer, able to reach speeds of 21mph for short periods.

Rorquals use the flippers on the sides of the body to steer. The flippers of the Humpback are especially large and have a jagged leading edge. All the whales have a small dorsal (back) fin to the rear of the body.

Other body features include a ridge between the snout and the blowholes (the whale's nostrils). The whale breathes out of the blowholes after surfacing, sending a spout of spray into the air.

BREEDING CYCLE
Rorquals are found in all major oceans of the world, the Atlantic, Pacific and Indian, in both Northern and Southern hemispheres. Most species spend the summer feeding in polar waters, where plankton and fish abound. They migrate south from the Northern hemisphere or north from the Southern hemisphere to winter in warmer waters. There they mate.

Males and pregnant females return to their feeding grounds for the summer, and the cycle begins again. When next they return to the warm water breeding grounds, the pregnant females give birth to their calves (young). They usually nurse their calves for 6 or 7 months, by which time they are back feeding again.

HAUNTING MELODY
As well as being a skilful acrobat, the Humpback whale is a fine singer. Most whales communicate with each other by sound – but while the others squeak and grunt, the Humpback sings a haunting melody.

The Humpback's song is made up of six basic themes, repeated over and over again. Each song can last for up to 35 minutes and may form part of a much longer recital. The variety of notes the animal uses have been described as resembling snores, whos, yups, chirps, ees and oos.

The whales' songs are studied using hydrophones (underwater microphones). Scientists have found that all the whales within a certain region of ocean sing much the same song. But the song changes according to the region and the season. The animals sing mainly when in shallow coastal waters and can keep in touch with one another over distances of more than 115 miles.

▶Underwater photographs of rorquals, such as these Humpback whales, show that the body is sleek and pointed towards the snout and not baggy as previously thought.

RIGHT WHALES

The Right whales are gathering in their breeding-grounds around the Bay of Fundy. Several can be seen travelling slowly on the surface. Over the still air come strange bellowing sounds like the lowing of cattle. The whales are calling to one another.

The Right whale was so called by the early whalers. It was the "right" whale to hunt because it yielded much oil and baleen (whalebone) and remained afloat when killed. And it was hunted, along with the Bowhead, almost to extinction. Despite protection, stocks of both whales are dangerously low, with a few thousand left.

The Right whale is much more vocal than the other two species of the right whale family, the Bowhead and Pygmy right whales. Both the Right whale and the Bowhead have an enormous mouth and long baleen plates. They use these to filter their food from the sea, like all the baleen whales. But they feed mainly on the surface, as does the Pygmy right whale.

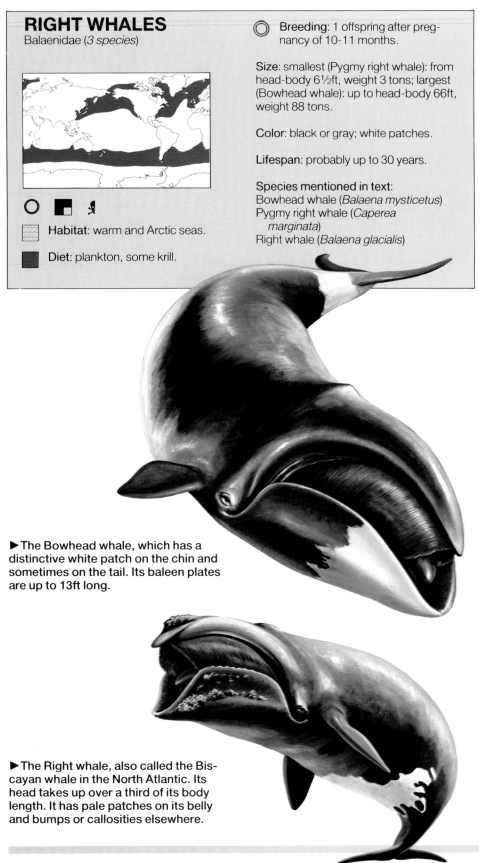

►The Bowhead whale, which has a distinctive white patch on the chin and sometimes on the tail. Its baleen plates are up to 13ft long.

►The Right whale, also called the Biscayan whale in the North Atlantic. Its head takes up over a third of its body length. It has pale patches on its belly and bumps or callosities elsewhere.

▲A southern Right whale and her calf in the warm shallow waters off Argentina. On the beach are basking elephant seals. In this species of whale, mating often takes place in waters 20ft deep.

▲A Right whale "lob tails," bringing its tail crashing down on the surface, before it dives. No one knows why it does this – probably it is just for fun.

▼This Right whale is feeding at the surface in water rich in plankton. The food is strained through the long baleen plates on its upper jaw.

BUMPS AND BARNACLES

Both the Right and the Bowhead whales have black skin marked with white patches. But the Right whale can easily be recognized by the bumps or outgrowths on its skin (one of these, known as the bonnet, is always present on the top of the whale's head). The Bowhead has none, nor has the gray-skinned Pygmy right whale.

These outgrowths, called callosities, are covered in barnacles and lice. They are usually found on the snout, under the jaw, over the eye, along the bottom lip and around the blowholes. The pattern of callosities is unique to each animal and so helps scientists to recognize individuals from some distance away.

WHALE POPULATIONS

There are several groups or populations of Right whales in both the Northern and Southern hemispheres. They live in warm or cool temperate waters. One population migrates between the Carolinas (winter) and Newfoundland (summer) off the eastern, Atlantic coast of North America. Southern populations winter around South America, South Africa, Australia and New Zealand, spending the summer feeding near Antarctica.

Bowheads spend most of their time in Arctic waters. One of the biggest populations feeds in the Beaufort Sea in summer and migrates south to the Bering Sea for winter.

In common with most other whales the Right and Bowhead whales tend to follow a 2-year breeding cycle that fits in with the north-south migration. Mating and calving usually take place in the warmer winter waters.

Not much is known about the breeding habits of the Pygmy right whale, which lives in the temperate seas of the Southern Ocean.

GLOSSARY

Adaptation Features of an animal's body or life-style that suit it to its environment.

Aggression Behavior in which one animal attacks or threatens another.

Air sac A fleshy pouch in the throat of some primates, such as the orang-utan, which helps magnify the sound of its call.

Antarctic The bitterly cold region in the far south of the world, around the South Pole.

Aquatic Living for much, if not all, of the time in the water.

Arboreal Living for much, if not all, of the time in the trees. *Compare* terrestrial.

Arctic The bitterly cold region in the far north of the world, around the North Pole.

Baleen The horny plates that grow in the upper jaw of some whales, used to filter food from the sea; commonly known as whalebone.

Bipedal Being able to walk on two legs, as in humans, rather than on four legs (quadruped).

Blowhole The opening of a whale's nostrils on top of its head, through which it "blows" air.

Blubber The thick layer of fat beneath a whale's skin.

Breaching The activity when a whale leaps out of the water.

Brindled Having dark streaks on a gray or tawny colored background.

Camouflage Color and patterns on an animal's coat that allow it to blend in with its surroundings.

Cannibalism The eating by an animal of others of its own species.

Canopy The upper layer of forest formed by the intermingling of branches and leaves. It may be closed (continuous) or open (with gaps).

Carnivore An animal that eats meat.

Cebids Monkeys of the family Cebidae, which include the capuchin, Squirrel and Night monkeys.

Cetacean A member of the animal order Cetacea, which includes whales.

Cheek pouches Pouches in the cheeks of Old World monkeys, used for storing food.

Class The division of animal classification above Order.

Competition The contest between two or more species over such things as space and food.

Coniferous forest Forest of trees with cones and needle-like leaves, which usually bear leaves all year.

Conservation Preserving and protecting living things, their habitat and the environment in general.

Copepods Small crustaceans that are a major food for the Baleen whales.

Crustaceans Shelled creatures, like shrimps, krill and copepods.

Deciduous forest Forest of trees that shed their leaves seasonally, usually in winter.

Den A shelter in which an animal or group of animals sleeps, hides, or gives birth to young.

Diet The food an animal eats.

Digit A finger or a toe.

Diurnal Active during the day.

Dominant Of higher rank; a dominant male of a group is "the boss".

Dormant Resting or asleep.

Dorsal Along the back, as in dorsal fin.

Echo-location The method bats and other animals use for finding food and their way around. They send out high-pitched sound waves and listen for echoes reflected from objects in their path.

Edentates An order of mammals which includes the anteaters and armadillos. The word Edentate literally means toothless, though only the anteaters are completely lacking in teeth.

Endangered species One whose numbers have dropped so low that it is in danger of becoming extinct.

Environment The surroundings, of a particular species, or the world about us in general.

Extinction The complete loss of a species, either locally or on the Earth.

Exudate Sap or gum that flows from a tree or other plant when its bark or stem is cut.

Family The division of animal classification below Order and above Genus.

Filter feeding The method Baleen whales take their food. They strain it through plates of baleen (whalebone).

Flukes The tail fins of a whale.

Generalist An animal that does not have a specialist life-style; one for example that can exist on a variety of foods.

Genus The division of animal classification below Family and above Species.

Gestation The period of time during which an animal is pregnant.

Glands Special organs that release chemicals on to the skin (scents) or in the body (hormones).

Habitat The kind of surroundings in which an animal lives.

Harem An animal group containing a single male and two or more females.

Helper An animal in a group that helps to raise young that are not its own.

Hibernation Period of inactivity during the winter cold.

Home range The area in which an animal usually lives and feeds.

Incubation Period during which an animal keeps an egg warm, allowing the embryo inside to grow.

Infanticide The killing of infants, which may occur, for example, when groups of chimps attack one another.

Insectivore An insect-eater.

Knuckle-walking Walking on all fours, using the knuckles of the forelimbs; practised only by chimpanzees and gorillas.

Krill Shrimp-like creatures, which form the main food of Baleen whales.

Lesser apes The gibbons and siamang.

Lob-tailing The beating of its tail on the water by a whale.

Lower primate One of the primitive primates known as the prosimians.

Mammals A class of animals whose females have mammary glands, which produce milk on which they feed their young.

Marine Living in the sea.

Marsupials An order of mammals, whose females give birth to very under-developed young and then raise them (usually) in a pouch. The echidnas, platypus and possums are marsupials.

Migration The long-distance movement of animals, usually seasonal, for the purposes of feeding or breeding.

Monotremes The echidnas and the platypus, so called because they have a single opening at the rear of their body. They are unique among mammals in laying eggs.

Nocturnal Active during the night.

Omnivore An animal that has a varied diet, including both plants and animals.

Order The division of animal classification below Class and above Family.

Plankton Small creatures that live in the sea.

Pod Term for a group of whales, occasionally of other animals.

Population A separate group of animals of the same species.

Predator An animal that hunts live prey.

Pregnancy Period during which the young grows inside the body of a mammal.

Prehensile tail One that can grip; some primates and marsupials have a prehensile tail.

Primates The order of animals that includes monkeys, apes and humans.

Prosimians Primitive primates that came before the simians – monkeys and apes – in order of evolution.

Rain forest Tropical and sub-tropical forest which has plentiful rainfall all year round.

Savannah The tropical grassland of Africa, Central and South America and Australia.

Scent marking Marking territory by smearing objects with scent from glands in the body or by means of urine.

Simians Monkeys and apes; higher primates.

Solitary Living alone for most of the time.

Species The division of animal classification below Genus; a group of animals of the same structure which can breed with one another.

Steppe The temperate grassy plains of Eurasia. Called "prairie" in North America.

Terrestrial Spending most of the time on the ground. *Compare* arboreal.

Territory The area in which an animal or group of animals lives and defends against intruders.

INDEX

Scientific names

The first name of each double-barrel *Latin* name refers to the *Genus*, the second to the *species*. Single names not in italic refer to a family or sub-family and are cross-referenced to the Common name index.

FURTHER READING

Alexander, R. McNeill (ed)(1986), *The Encyclopedia of Animal Biology*, Facts on File, New York

Berry, R.J. and Hallam, A. (eds)(1986) *The Encyclopedia of Animal Evolution*, Facts on File, New York

Corbet, G.B. and Hill, J.E. (1980), *A World List of Mammalian Species*, British Museum and Cornell University Press, London and Ithaca, NY

Griffiths, M.E. (1978), *The Biology of Monotremes*, Academic Press, New York

Hunsaker II, D. (ed)(1977), *The Biology of Marsupials*, Academic Press, New York

Jolly, A. (1972), *The Evolution of Primate Behavior*, Macmillan, New York

Kingdon, J. (1971-82), *East African Mammals*, vols I-III, Academic Press, New York

Macdonald, D. (ed)(1984). *The Encyclopedia of Mammals*, Facts on File, New York

Moore, P.D. (ed)(1986), *The Encyclopedia of Animal Ecology*, Facts on File, New York

Moynihan, M. (1976), *The New World Primates*, Princeton University Press, Princeton

Nowak, R.M. and Paradiso, J.L. (eds)(1983) *Walker's Mammals of the World* (4th edn) 2 vols, Johns Hopkins University Press, Baltimore and London

Slater, P.J.B. (ed)(1986), *The Encyclopedia of Animal Behavior*, Facts on File, New York

Wimsatt, W.A. (ed)(1970, 1977), *Biology of Bats*, vols 1-3, Academic Press, New York

Winn, H.E. and Olla, B.L. (1979), *The Behavior of Marine Mammals*, vol 3, Cetaceans, Plenum, New York

ACKNOWLEDGMENTS

Picture credits

Key: *t* top *b* bottom *c* centre *l* left *r* right
Abbreviations: A Ardea. AH Andrew Henley. AN Nature, Agence Photographique. ANT Australasian Nature Transparencies. BC Bruce Coleman Ltd. J Jacana. NHPA Natural History Photographic Agency. OSF Oxford Scientific Films. SA Survival Anglia Ltd.

6 A. 8 Natural Science Photos. 9 J. Visser. 11 A. 16/17 A. 18 Andy Young. 20 A. 21*tl* Rod Williams. 21*r* C. Jenson. 23 AN. 24 A. 28 Dawn Starin. 29 Rod Williams. 30 J. MacKinnon. 31*l* Anthro-Photo, 31*r* A. 32*l* AH, 32*r* J. 33 J. MacKinnon. 34 BC. 35 AH. 36 J. 37 J. MacKinnon. 38*t* A, 38*b* P. Veit. 39 P. Veit. 41 NHPA. 43*t* J. 44 J. Payne. 46/47 BC. 48 Eric and David Hosking. 49*t* BC, 49*b* BC. 50 A. 51 D.R. Kuhn. 52 OSF. 53*t* M. Fogden, 53*b* A. 55*t* J, 55*b* J. 56*t* A. 56/57 BC. 58 OSF. 59 NHPA. 60 A. 62 WWF/S. Yorath. 63*tl* ANT, 63*bl* ANT, 63*r* J. 64 BC. 65*t* Frithfoto, 65*b* AH. 66 E.

Beaton. 67 G. Mazza. 70 OSF. 70/71 OSF. 72 AH. 73 ANT. 74 AH. 76 SA. 77 BC/M.P. Price. 79 P. Morris 80*l* A, 80*r* BC. 82 A. 83 AN. 85 K. Balcomb. 86/87 PB/James Hudnall. 88 SA/J. & D. Bartlett. 89*t* W.N. Bonner, 89*b* BC.

Artwork credits

Key: *t* top *b* bottom *c* centre *l* left *r* right
Abbreviations: PB Priscilla Barrett JC Jeanne Colville DO Denys Ovenden MM Malcolm McGregor

7 PB. 8 JC. 9 PB. 10 PB. 11 PB. 13 PB. 15 PB. 18/19 PB. 19 PB. 20 PB. 22 PB. 25 PB. 26/27 PB. 29 DO. 33 PB. 35 PB. 36 PB. 39 PB. 40/41 DO. 42 DO. 45 DO. 49 PB. 51 DO. 54 JC. 57 DO. 60 Graham Allen. 61 Graham Allen. 64 JC. 66 JC. 68/69 Dick Twinney. 72/73 PB. 74 Dick Twinney. 75 Dick Twinney. 78/79 PB. 81 PB. 82 MM. 84/85 MM. 88 Rob van Assen.